JEREMIE AVEROUS

With Thierry LINARES

Advanced Scheduling Handbook for Project Managers

JEREMIE AVEROUS
With Thierry LINARES

Advanced Scheduling Handbook for Project Managers

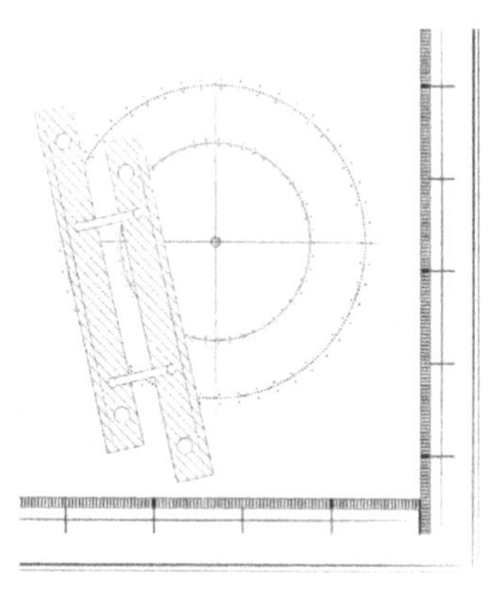

A Practical Navigation Guide on Large, Complex Projects

First Edition - 2015

Published by Fourth Revolution Publishing, Singapore
A trademark of Fourth Revolution Pte Ltd
23D Charlton Lane, Singapore 539690,
www.FourthRevolutionPublishing.com

In this book, the alternating of gender in grammar is utilized. Any masculine reference shall also apply to females and any feminine reference shall also apply to males.

ISBN: 978-981-09-4322-6

ISBN e-book: 978-981-09-4323-3

Published in Singapore.

First print – Print-on-Demand, July 2015 / worldwide availability on all e-bookshops through LightningSource

We can Customize this Handbook for your Organization

We have found that organizations find great value to have their own internal handbooks to cover Project Control issues. These full-color, custom-branded handbooks can be spread to a wide population.

Thanks to our publishing partnership, this generic handbook has been successfully customized and printed in many hundreds of copies for some major global Project-driven businesses.

Conducted after a careful review of the organization's particularities, and tailored to respond to your needs, such a customization includes:

- Specific systems and processes,
- Specifics of the industry and construction context.

Contact us for a customization Project.

Get Advanced Training delivered

The best practice concepts of this handbook get best spread in your organization through training programs. We deliver trainings globally based on the contents of this handbook, and customized to your particular needs. We are recognized as Registered Education Provider by the PMI®.

Contact us for a training Project to bring your organization to the next level in terms of Project Scheduling.

Contact

Contact @ ProjectValueDelivery.com

We Empower Organizations to be Reliably Successful in Executing Large, Complex projects.

Discover more on

www.ProjectValueDelivery.com

In the Same Collection

1. The Pocket Guide for those Daring Enough to Take Responsibility for Large, Complex Projects
   by Jean-Pierre Capron

2. Project Soft Power, *Learn the Secrets of the Great Project Leaders*
   by Jeremie Averous

3. Practical Cost Control Handbook for Project Managers, *a Practical Guide to Enable Consistent and Predictable Project Forecasting*
   by Jeremie Averous

4. Practical Project Risk Handbook for Project Managers, *a Guide to Enhance Opportunities and Manage Risks on Projects*
   by Jeremie Averous

5. Advanced Scheduling Handbook for Project Managers, *a Practical Navigation Guide on Large, Complex Projects*
   by Jeremie Averous with Thierry Linares

6. Practical Project Control Manager Handbook, *from Back-Office Manager to Project Strategist*
   by Jeremie Averous *(to be published late 2015)*

And more to come…

Discover the latest publications and more on:

www.ProjectValueDelivery.com

Contents

Foreword by Jean-Marc Aubry

In the Oil and Gas business as in industrial projects, the shareholders, investors and Clients are increasingly requesting fast-track schedules and fast completion of Projects. The objectives are to optimise their financing costs and conditions, satisfy the market and their own Clients, benefit of competitive advantages, help them to reduce the pay-back as much as possible and keep their leading positions.

Today, all Parties including Contractors are facing very difficult schedules and related complex strategies to meet these objectives. The risk-taking has increased, the floats are disappearing and any delay is a disaster for the Contractor, his direct Client and the shareholders.

Still around 20 years ago, we were working on sequenced phases: engineering, procurement, construction, start- up and Operations (EPC), where nowadays people are more and more dreaming of CPE...

I like to take the story of the driver leaving for a long journey by car. Either he jumps in the car, with the map on the side seat and tries regularly to look at his way, or he spends a bit of time to study ahead the best route, combining estimated time, distance, highways and gasoline costs, safety aspects, anticipating traffic jams... Both will arrive at their final destination. The first one with no control on his journey and constant pressure, the second one will arrive on time and at the best satisfaction and safety of his passengers.

This illustrates perfectly what scheduling and planning are for a Project.

For some people, the schedule is a thick Primavera document, established by the Scheduling department, and that nobody understands or looks at.

For the experienced Project Director, it will be the perfect translation of his Project strategy. It will contain his analysis of the risks and the corresponding mitigations, the floats he can anticipate, the contractual approach he has to implement, the key milestones he has to meet to secure step by step his Project schedule-wise but also budget-wise.

Out of a full booklet of detailed schedules, he will be able to extract a simple document, with key milestones and critical paths, which will help him drive his Project to the right end, anticipating any disturbance by the proactive actions he will be able to implement.

To establish such a route book, the Project Director counts on his most critical partner who is the Schedule Manager, able to understand the strategy of the project, the risks, and develop, thanks to a large experience in this 'Science', the right itinerary.

This partner will also be the 'Master of the Temple', alerting, simulating, evaluating alternatives, to continuously help to stay on the road.

The authors have become this kind of Master, building a very rich experience in this crucial domain, continuously looking for improvement and creative approaches, which are the key of a successful Project.

I have been very honoured to write a few words for this handbook, our experience on the Yamal Project is not only a professional journey, but also a personal and high value cooperation based on trust. Very few people have developed a so deep understanding of their role and mission. This may be what we call simply professionalism.

Thanks for this very interesting analysis which will benefit to the profession.

Jean-Marc AUBRY
Yamal Project – Fellow Executive Project Director
Previously President, Technip France

Introduction to Project Navigation

Why Project Navigation?

A Project is like an intercontinental sailing expedition. When executing a Project, one needs to define the goal and a plan to reach it; and then, fit out a vessel with the appropriate quantities of fuel and supplies to last for the long voyage, and finally, recruit the right crew.

The days are long gone of the adventurers that cast off without any idea of goal or direction – that may still exist in some R&D contexts but very rarely in the context of Large, Complex construction Projects.

When the vessel has finally left the shore to begin the uncertain voyage towards a new continent, left to the forces of the sea, the currents and the winds, there are three fundamental navigation questions that require an accurate response, on a regular basis:

- Where are we?
- Where are we heading to (if we continue according to the present trend)?
- What adjustments do we need to do to come back on course?

The result of these three questions leads to a decision that needs to be taken consciously - in the present circumstances, whether or not to amend the current course of the vessel, or the sails' configuration. Proper navigation and decision-making are intertwined.

Our experience in consulting for Large, Complex Projects shows that a large percentage of the Project organizations –half of them maybe– cannot even respond adequately to the first question ('where are we'). They fail to have adequate monitoring tools and information gathering processes that would give them an accurate picture of the actual status of the Project. As a result, they take navigation decisions on the basis of inaccurate status information!

> Our experience shows that many Large Projects don't know exactly where they are; and in the remaining projects, a large proportion can't fully forecast where they are going.

Then, amongst those other organizations that succeed in maintaining accurate status data, another large proportion are not able to identify and extrapolate the observed trends and determine how much they will deviate from the initial navigation plan if they were to continue unimpeded. Even more so, when seeing a storm form on the horizon, many Project Managers today are at a loss to take the right decisions to react to it by changing course or changing the sails' configuration. This lack of capability for anticipation has thrown many sailors helpless on chartered reefs, and continues to do so in the realm of Project management.

Today, automated systems on ships and aircrafts still do constantly implement small course corrections based on the same three questions so as to reach safely their destination. These automated systems are able to respond to small perturbations but human monitoring is still required for the management of large and unexpected changes. Actually, the issue today is still to teach Captains the basics of good ol' navigation so that they understand what is important and where and how automated systems can support - and where they find their limits.

In the field of Project Management, we observe that Project Managers today have sometimes taken a back seat behind those terrific scheduling systems that seem to promise full control on all events happening on the Project, like a GPS. This is an illusion – those systems are generally not well implemented, and not properly used. They have intrinsic limitations that are not well understood. Most successful Project Managers today still fly by the seat of their pants and smell the sea to navigate prudently. However these experienced hands that started their career scheduling manually on paper will soon be part of the heroic history of Project Management, and the newer generations seem to have lost the feeling for the reality of actual Project navigation.

It is the ambition of this handbook to re-establish the basics lost in time of Project Scheduling for today's Project Managers. Automated tools can have great power when one knows how to use them properly and understands their limits, and they should be used to leverage what can be achieved – but in no case can they replace the sound judgment and experience of Project Managers.

There lies the ambition of this handbook – a Navigation Handbook for the Project Managers of Large and Complex Projects of modern times.

A Practical Handbook

This handbook has been specifically written in the particular context of Large and Complex construction Projects.

This handbook does not intend to be a 'starters' guide to Project scheduling as it tackles advanced topics. It does not cover, in particular, the basic technicalities of scheduling, but concentrates on the perspective of Project Management's check and usage of schedule as a decision-making tool. In other words, it concentrates on how the Captain should use the navigation process to reach the destination safely – but will not describe the sometimes tedious navigation calculations.

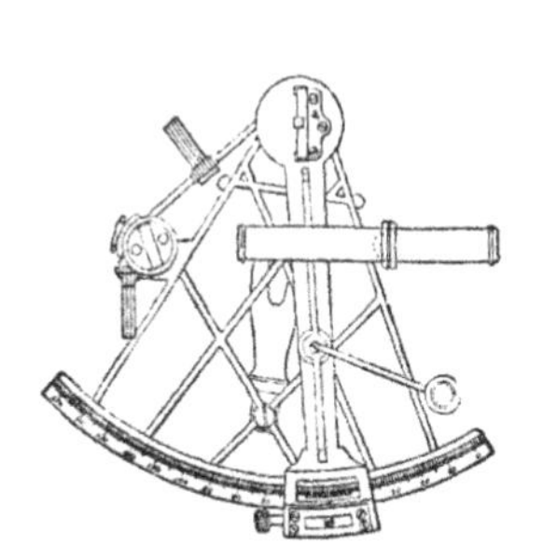

We will use historical sailing ship navigation instrument drawing such as the one on the left to highlight those sections that refer to the principles of navigation.

Those sections will form a thread running throughout the entire book, where traditional navigation tools and processes will be quoted as a useful metaphor to put the concepts in perspective.

Our intent is that well worn-out copies of this navigation handbook will be found on every Planner's, Project Control Manager's and Project Manager's desk.

The Handbook's Structure

Chapter 1 covers a number of sound Scheduling Golden Rules that summarize key principles that are essential to proper Project schedule management. All those concepts are developed in this handbook.

Chapter 2 describes the best practice in terms of the set of different schedules (the '*schedule hierarchy*') needed to support the execution of Projects, for any size or complexity. Essential pieces are the Convergence Plan and the Integrated Project Schedule. Chapter 3 and 4 describe best practices for building these two central components.

Chapter 5 and Chapter 6 give key insights for Project Managers to check the quality of schedules and improve them prior to, and during Project execution. Chapter 7 covers in particular what insights can be drawn from Schedule Statistical Analysis.

Chapter 8 and 9 describes the best practices for the central scheduling processes during Project execution: updating and re-forecasting.

Chapters 10 to 12 cover advanced topics related to Project scheduling:

- Ensuring sufficient agility of the Project Schedule to face the inevitable changes,
- Introducing the concept of schedule buffer as a way to create more realistic and manageable Project schedules,
- Explaining the basics of how to use the schedule for contractual purposes, in particular in cases of request for Extension of Time.

Finally, Chapter 13 summarizes how to assess effectively the quality of the Project Scheduling process. The Chapter contains a number of easy-to-use reference checklists for Project reviews.

Topics Not Covered in this Handbook

This handbook is focused on the practice of Project Scheduling during Project execution. The following topics are not covered in this handbook:

- Technical use of scheduling tools/software,
- Basic scheduling concepts,
- Quantity surveying,
- Duration estimating (at studies phase or during Project execution for changes).

Who Is This Handbook For?

This handbook is explicitly for Project Sponsors, Project Managers, Project Control personnel (in particular Planning Managers and Leads) and all those who aspire to become Project Managers; Budget Owners within Projects (Package Managers, etc.) as well as functional managers that are involved in scheduling and resource planning.

This handbook has been primarily written from the perspective of the Project Manager or Project Management Team, who use schedules to take decisions in action. It is quite different from the usual perspective of schedule professionals who are sometimes centred on the excellence of their tools' usage. This handbook thus differs markedly from most books on project scheduling. By taking this different point of view, we believe that this book will fill-in a much needed gap between Project management and Schedule professionals and create useful conversations in organizations.

Chapter 1: Scheduling Golden Rules

The main objective of Scheduling management is to enable the Project Manager and its management to take decisions derived from an accurate current knowledge and understanding of reality, with the aim of reaching a successful project outcome.

From this broad objective, a number of Golden Rules describe the basic requirements of Schedule Management.

In all instances, 20 Golden Rules need to be followed when it comes to Project Scheduling.

1. **Accountability:** Budget Owners are ultimately accountable for their schedule (including update and forecast). Planners support and challenge Budget Owners. The Project Manager is ultimately accountable for the entire Project schedule and shall dedicate sufficient time and effort on this essential navigation tool.
2. **Project Scope, Cost & Schedule consistency:** the Project schedule is at all times consistent, comprehensive and intrinsically linked with the two other sides of the Project Triangle: the Project Scope, and the Project Cost. The Project Scope is described in the main Contract or specification, including Change Orders and approved Changes. In particular, the *Project Cost Model* (including cost time-phasing consistent with the Project schedule) is continuously updated by the Project Control team consistently with the schedule.

3. **Align with the Project strategy:** in addition to remaining consistent with cost and scope, the schedule responds to the Project strategy: contractual strategy with the Owner/ Client and suppliers, decision-making logic, and more generally, execution strategy of the Project and key success drivers.
4. **Develop schedules from the top down**: to ensure alignment with the Project intent and design of best execution strategies, develop schedules from the Project strategy and objectives i.e. from the top of the scheduling hierarchy, and not from Detailed Functional Schedules.
5. **Reflect reality candidly:** The schedule must reflect candidly the reality of the Project progress status and associated re-forecast, however difficult or annoying this reality could be.
6. **Immediacy principle:** It is essential to reflect significant new schedule variances as soon as their occurrence is known (e.g. internal or Owner's instruction to proceed), at least in terms of order of magnitude, even if their exact final duration has not been fully assessed. Subsequently, immediate notification of the other Party to the contract is also essential to protect one's commercial interest.
7. **Implement a Proper Schedule hierarchy and formats.** Different scales, details level and views are suitable for different usages. Build a consistent schedule hierarchy and make good use of the different detail levels. Use different schedule views for different purposes and users.
8. **Limit detail and complication of the Integrated Project Schedule.** It should focus on functional interface and critical areas. 2,000 to 2,500 activities would be a maximum, with an emphasis on links between functions, and a proper balance between Project phases and functions. Choices will have to be made. It is linked to the necessity to have a comprehensive schedule hierarchy to respond to the needs of all Project contributors and stakeholders.

9. **Increase the schedule robustness and resilience instead of minimizing the Critical Path.** Increase the float of non-critical sequence of events to ensure they will not become critical, and introduce allowances and a contingency managed by the Project Manager.
10. **Float and Buffers are to be owned by the Project Manager.** Float knowledge and ownership should not be spread through the Project team relinquishing effective control. It is an essential Project management tool.
11. **Fight the 'virtual' float creation.** When a schedule moves to the right because of delays, in effect it creates float for all activities that have to be performed. Avoid this pernicious effect by sticking to the discipline of Convergence Planning and updating your Integrated Project Schedule so that this 'virtual' float is not unduly created where it should not. Introduce explicit buffers if required that remain under the control of the Project Manager.
12. **Be disciplined in updating the Convergence Plan.** Don't change the dates of the gates and only show deliverables completed when they are 100% complete. And when there are deviations, actual or forecast, the Project Manager and the supporting organization must take the relevant recovery actions.
13. **Update the schedule bottom-up** based on the Project extended team's knowledge.
14. **Check regularly the quality of the schedule update** to make sure decisions are taken on a robust basis.
15. **Base the schedule re-forecast on a root cause analysis,** not a simple bottom up approach. Use Earned Schedule as a useful challenge. Don't forget to reforecast accordingly future activities that would be impacted by the same root cause.
16. **Reforecast future activities based on the knowledge acquired from ongoing and past activities**. This is too often forgotten in schedule updates.

17. **Accuracy over precision:** schedule updates should be accurate but not necessarily precise. This important distinction should focus the effort of the scheduling team (ref. to Chapter 7).
18. **Ensure full traceability of all schedule logic changes.** This will help to support or defend against future claims.
19. **Raise Extension of Time requests as soon as they are known,** through the proper channel as described in the Contract. This will avoid cumbersome debates in hindsight and will ground compensation decisions in current reality.
20. **Understand and compensate for the known psychological biases at play in Project schedules.**

Major psychological factors at work in project schedules

- **Parkinson's law:** work tends to fill the time available [if the task if planned for longer than it would take, work will still take at least that duration]
- **Student's syndrome:** if people have time to do a task they will always wait for the last possible moment to start
- **Commitment syndrome:** people will always 'pad' their duration estimates when they are asked (consciously or unconsciously) to commit to a duration. Hence they announce durations that can be much longer than what is achievable.
- **Planning Fallacy:** an optimistic bias on the duration of own's future tasks, irrespective of benchmark data on past duration distribution of similar tasks (Daniel Kahneman, Amos Tversky).
- **Lack of calibration of estimates:** when people have not calibrated their estimates comparing to actual durations, they will tend to be pessimistic (conservative) for usual tasks and optimistic for unknown tasks.

Chapter 2: The Schedule Hierarchy

Chapter Key Points:

- Don't put all the details in a single schedule but implement a schedule hierarchy with various levels of details.
- The Integrated Project Schedule linking all E-P-C activities is the key piloting tool for the Project. It should not be too detailed and should be focused on the inter-relationships between functions.
- The Convergence Plan allows for a higher level control and communication with the team.
- Detailed schedules for specific functions or suppliers are required to have a sound baseline and track the details of the work.
- It can be required sometimes to operate with different schedules for different stakeholders. This should be avoided for the Integrated Project Schedule, and is permissible for lower level schedules within a strict framework.

Introduction

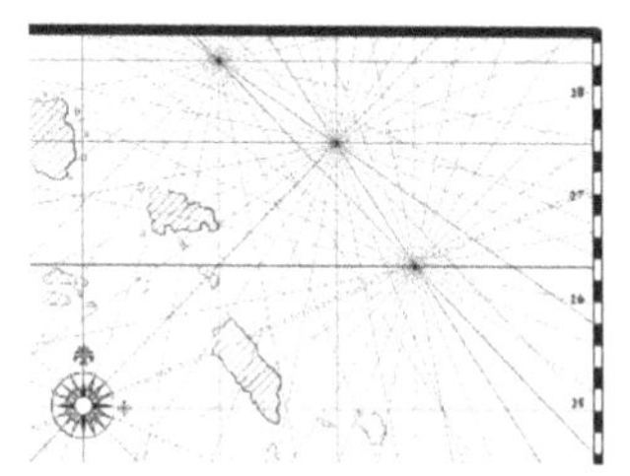

Project schedules are a bit like paper maps: depending on the usage you would not use the same scale and the same level of details. A 1/25,000 map is good for walking around but unusable for driving or flying; conversely a 1/10,000,000 map is good for looking at optimizing an inter-continental trajectory but not so much for local navigation.

With the scale comes a level of details that differs.

Oceangoing vessels are required to carry a full suite of nautical charts of various scales, complemented by books that give details as to navigational requirements and harbour details. Notices to Mariners, in addition, give updated information.

The Project Manager should focus on the overall route strategy and not spend time on the detailed Project maps, except when it comes to prepare a specifically sensitive situation such as a coast landing. Other participants to the Project should look into more details and should have higher resolution maps.

Like it is absurd to believe that a single map could be used for all phases of navigation, it is absurd to believe that a single schedule with the finest level of detail will do the job on a large Project – and for all contributors. This Chapter explains why and describes how to build a consistent hierarchy of schedules to keep control of the Project.

Recommended Schedule Hierarchy Overview

Different schedules are thus meant for different uses and different users. There are three main schedule levels:

- **Strategic level**, where the schedule provides an overview allowing long-term decisions and orientations: financing principles, contract placements, taskforce and key equipment localization and mobilization, etc.,
- **Project coordination level**, where the proper consolidation of available information from all project contributors (engineering capabilities, construction needs, external constraints, etc.) is performed. This allows logical Critical Path analysis, trending and reforecasting,
- **Operational level**, where the priorities decided by management are cascaded down in a detail that only the specific functions can have a real control on. This allows the management of the work at the lowest granularity. From those details, the operational level also allows to feed back up the

necessary statistical data for trends and reforecasting, as well as progress measurement.

Conventional Schedule Hierarchies

When executing a large and complex Project, a number of schedules are used with different levels of detail and scope. These different schedules need to remain consistent.

There are a number of more or less standardized schedule levels (ref. for example to AACE international's recommended practice):

- **Level 1:** high level schedule showing the major components and contractual milestones of the Project. It is generally a 1-2 pages schedule consisting of less than 100 activities,
- **Level 2:** a schedule that can be more detailed than the level 1, in particular regarding milestones and trade/ function-specific activities. Depending on the reference there might be a requirement for activity linkages and identification of a high level Critical Path, however generally this schedule is not logically linked. It is typically a schedule of less than 300 activities and is in effect rarely used by the Project manager. It can be used as a 'sign-off' schedule for contractual purposes,
- **Level 3: the Integrated Project Schedule:** a schedule that has a level of details and the adequate linkage so as to enable the identification of the Critical Path of the Project. It is used to monitor and control effectively the execution of the Project. It can be resource-loaded for specific critical resources. This allows for resource utilization analysis or resource-levelling if required,
- **Level 4-5: detailed work schedules** that show only one particular area of Project execution. They can be by trade/ function, work phase or by work area and are designed to be used directly by the site or function managers for their day-to-day execution planning, progress tracking and reporting. They should be resource-loaded for those critical resources to allow for resource utilization analysis or resource-levelling if required. They are not

necessarily developed using pure scheduling tools or logically linked, and can be shown in other systems e.g. document register.

This hierarchy of schedules is very general and does not relate to the usage made by the Project Manager of these schedules. Levels 1 and levels 2 are generally only used by senior management or for contractual purposes.

Appendix 5 explores in more detail the different schedule levels and how schedule levels need to be linked to the levels of the Project's Work Breakdown Structure (WBS).

Project Value Delivery's Recommended Project Execution Schedule Hierarchy

From the Project manager's perspective, during Project execution, the '**Integrated Project Schedule**' (level 3) is the key document. We discuss later the appropriate characteristics and level of detail of such schedule. However it is also important to keep a high level view of the entire Project schedule that can fit on one page, and ideally, shows a good representation of the main workflows, where they are supposed to converge, and of the most critical deliverables. This is the '**Convergence Plan'**, which has the advantage on being a better-adapted tool for complex projects' management than simple higher-level roll-up schedules.

The actual hierarchy of schedules recommended by Project Value Delivery to be used by the Project Manager during execution is thus the following:

- **Convergence Plan** – a graphical representation of the Project workstreams, convergence points, and key deliverables. It does not include linkages. It serves as a high level overview and is updated regarding progress. It is posted in the Project area for all to see and to monitor as part of Visual Management. On a large Project the Convergence Plan contains around 100 to 150 key deliverables for the entire Project,
- **Integrated Project Schedule** – a fully linked schedule covering the entire Project execution scope with an adequate level of detail to take decisions at

the Project Management Team level and identify the Project drivers. It also remains sufficiently high level to still allow agility to cater for changes due to unexpected circumstances and ensure proper, accurate update and forecasting on a monthly basis. In practical terms, this means that the Integrated Project Schedule should aim to have a size of 2,000 to 2,500 activities (without counting milestones etc.)
,

- **Simplified Project Schedule** – a fully-linked schedule covering the entire Project execution scope with 200 to 400 activities, that covers the Project's main drivers and interface points. It is not a level 2 schedule as per the usual definition because it is fully logically linked, and for that reason cannot also be just a summarized view of the Integrated Project Schedule. It is mainly used for scenario planning and Schedule Statistical Analysis. It may also be used as a support for Extension of Time negotiations at management level. It is maintained in parallel to the Integrated Project Schedule during the course of the Project,
- **Detailed Functional Schedules** – schedules or other tools involving delivery dates, that do not cover the entire Project scope but present much more detail as to the delivery of those scopes, down to the elementary deliverable as appropriate. They should be resource-loaded at least for critical resources to allow for resource utilization analysis or resource-levelling if required.

This hierarchy of schedules is summarized in the figure on the next page.

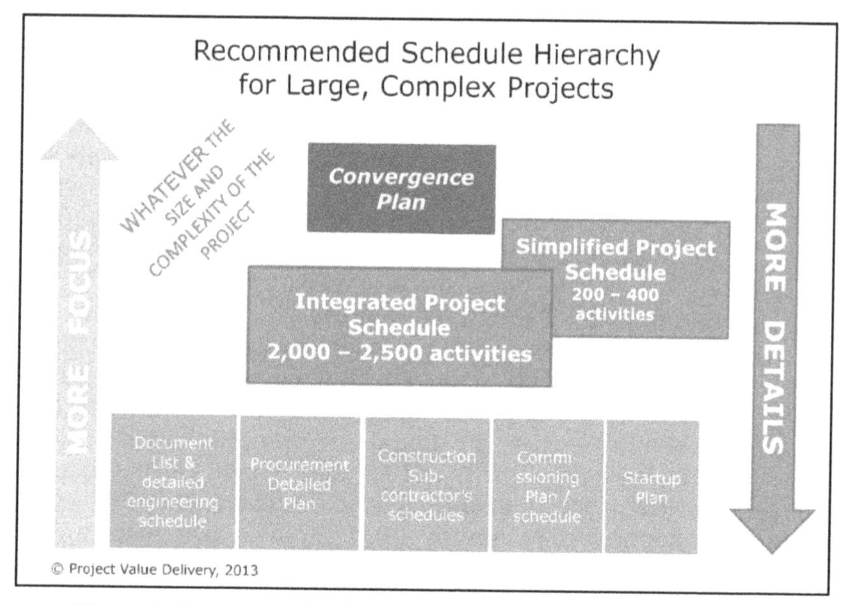

Figure 1: Recommended Schedule Hierarchy for Large, Complex Projects

Conventional levels	PVD's recommendation	Benefits
Level 1	Convergence Plan	A tool focusing on convergence issues rather than just a roll-up with activities covering the entire Project.
Level 2	Simplified Schedule	Contrary to usual "roll-up" level 2, a fully linked simplified schedule
Level 3	Integrated Project Schedule	Limit the size of the level 3 schedule to make it more accurate and agile
Level 4	Detailed Functional Schedules	Detailed schedule per function with substantial additional information compared to the Integrated Project Schedule.
Level 5	Further detailed schedules (as required)	Detail down to the single deliverable or basic task

Functional Usage of the Different Schedules

The different schedules in the schedule hierarchy respond to different functional needs. The table below summarizes this functional analysis.

	Usage for PMT	Internal Communication	External Communication
Convergence Plan	Project strategic planning, float monitoring, prioritisation. Key decision points.	Overview communication to team.	Overview communication to Senior Management.
Simplified Schedule	Schedule Statistical Analysis. Scenario analysis at PMT level.	Overall mobilization plan by operational center	Extension of Time explanation and presentation. Scenario analysis. Possible contractual schedule
Integrated Project Schedule	Progress monitoring. Critical Path identification. Float measurement and monitoring. Interface between functions and with stakeholders. Decision points. Schedule forecast.	Progress monitoring. Critical Activities identification. Interface between functions. Schedule Forecast.	Progress monitoring. Interface with free-issued items / stakeholders Schedule Forecast Extension of Time analysis and justification
Detailed schedules (1)	Support to analysis. Resources review and planning.	Detailed work assignments within the functions. Resource review and planning.	*Not applicable (except to support forensic analysis)*

(1) Detailed schedules can also be the Integrated Schedules of individual contractors and suppliers.

The Convergence Plan

In Projects there are general a few critical convergence points where many deliverables, parts and equipment must meet in a single location at the same time. In Complex Projects, failed convergence (the unavailability of one or several key deliverables at the moment required at a key convergence point) is the most common source of dramatic consequences on the Project outcome, in terms of delays and cost overrun. It needs to really be avoided because of the disruption it creates. Sometimes, the unavailability of very simple parts can delay a billion dollar Project by several weeks. The Project leader and his team needs to be aware of this effect and focus on avoiding such an occurrence in particular if it is unanticipated.

The Convergence Plan should be the first plan to be established upon mobilization of the Project, within the first few days (with the proviso that exact dates can be finalized later after the Integrated Project Schedule has been approved). Its development will allow a meaningful conversation amongst the newly mobilized Project team about Project drivers, allow identification of key missing items in the plan, and otherwise help to drive the first critical activities until such time that the baseline Integrated Project Schedule is established (which will take up to a few months on a large Project, depending on the maturity of the schedule prepared at tendering/ feasibility stage).

The Convergence Plan is a proven key tool for the Project Management Team to identify and monitor the critical deliverables and convergence points of the Project. It has been developed initially in the automotive industry where it has been proven to ensure timely delivery of Projects – provided discipline is maintained in its implementation.

To achieve a proper convergence planning, the following logical steps need to be followed:

- Prepare a meaningful Convergence Plan:
    - Identify the key workstreams of the Project and where they converge,
    - Identify what are the key deliverables that are required for Project delivery (those without which the Project could not be delivered) and put them on a timeline, focusing on those early deliverables that are required in the first half of the Project,
    - Group these deliverables around critical convergence points of the Project,
- Monitor and take action on deviations:

- Monitor regularly the effective convergence of all those deliverables,
- Take massive actual action in case of deviation that might become critical.

It looks simple but is not so easy. In Chapter 3 we detail some practical recommendations to ensure the success of this process, both for at planning stage and for its ongoing utilization.

The Integrated Project Schedule

The Integrated Project Schedule is the schedule describing the entire Project scope. It is used by the Project Manager for forecasting and decision-making.

The Integrated Project Schedule needs to:

- Be fully linked with proper sequencing of activities and with minimum artificial constraints so as to enable determination of the Critical Path and ensure that actual progress (delays, completion in advance) will transfer into the network of future activities and impact all dates including the Project final delivery date,
- Include activities from all trades/ functions in a just sufficiently detailed manner to drive these trades' more detailed planning; and show a balance between all the Project functions such as engineering, procurement, construction, commissioning,
- Be adequately weighted for proper progress monitoring,
- Be adequately resourced for resources critical to the Project,
- Be updated to reflect at all times (or, at least once a month) the latest status of physical progress and the reforecast of the current and future activities,
- Enable changes to reflect actual execution changes, as well as scenario analysis to support decision-making.

Chapter 4 describes in detail how to develop an Integrated Project Schedule. The main highlights are:

- Ensure that the Integrated Project Schedule responds to the needs of Project execution and is aligned with the Project goals and execution strategy,
- Limit the size of the Integrated Project Schedule so as to keep it manageable for updates and changes,
- Ensure the Integrated Project Schedule is properly focused on the interfaces between the functions and is balanced between them,
- Ensure the Integrated Project Schedule is properly coded to allow Earned Value Management when appropriate, as well as to produce the different views needed by different Project contributors,
- Ensure the Integrated Project Schedule is properly resourced when resource availability is a constraint for Project execution.

The Simplified Project Schedule

One of our key recommendations is to produce a representative Simplified Project Schedule and update it regularly during Project execution. This schedule needs to be maintained entirely consistent, at all times, with the Integrated Project Schedule.

Producing the Simplified Project Schedule

It takes time and effort to produce a meaningful Simplified Project Schedule. Because it needs to be logically linked, and avoiding complicated Start-to-Finish links or other planner's tricks so as to be usable for Schedule Statistical Analysis, it cannot be just a filtered or rolled-up higher-level view of the Integrated Project Schedule.

It cannot either be just limited to a representation of the (current) Critical Path, because the goal is to identify those other near-critical or currently remotely critical chains of activities that might suddenly drive the entire schedule.

Because it requires an in-depth understanding of the Project's actual drivers, producing the Simplified Project Schedule cannot be left to the Project planners alone. It requires the Project Manager's involvement and engagement, together with the key Project team members, preferably in a workshop. The exercise will actually be found to be very worthwhile at the start of the Project as this cognitive exercise will allow the team to have a deeper understanding of the success factors for the Project.

Maintaining the Simplified Project Schedule

The Simplified Project Schedule once produced can be updated regularly by the planning team, consistently with the latest Integrated Project Schedule update. Because the Simplified Project Schedule only consists of 200 to 400 activities, which might not be activities or groups of activities of the Integrated Project Schedule, the update is best done manually.

Using the Simplified Project Schedule

The Simplified Project Schedule is an asset for Project management. It is used to run Schedule Statistical Analysis that makes sense whenever needed (ref. Chapter 7). It can be used for quick scenario planning when issues arise that might require a re-baselining of the Project execution plan. It can be used to communicate on the reasons for a request for Extension of Time (ref. Chapter 12). Finally, it can be used as a communication tool with some stakeholders, in particular:

- The organization's management, who might even find useful to upload the Simplified Project Schedules of various Projects together as part of a consolidated portfolio or programme management schedule,
- Other parties to the Project who can use it as an easier way to understand the unfolding of Project events (authorities, lenders etc.).

The Lower-Level Functional Schedules

The lower level detailed functional schedules cover only a part of the scope, in a very detailed manner. They can be schedules linked logically or simply a set of activities or deliverables with associated dates (typically called a register).

Typical lower-level schedules include:

- An Engineering Master Document Register including the comprehensive list of documents to be produced, the associated forecast and actual dates (in large engineering organizations there can even be a fully linked detailed engineering schedule with all relevant documents and revisions, in particular when the engineering process is complicated, iterative and inter-related such as in process engineering),
- A Procurement plan including a comprehensive list of purchase orders and contracts for services, including associated bid, award and delivery dates,
- A Logistics plan (for remote sites) including logistics arrangement dates,
- A Construction schedule (possibly subdivided by area) detailing the construction activity details,
- A Commissioning schedule (possibly subdivided by systems) describing the detailed pre-commissioning and commissioning activities.

These detailed plans are often rolling plans with a high detail and accuracy of the coming weeks/ months and lesser detail and accuracy of later activities. They are generally produced and maintained by the personnel responsible for that particular area or function, who also directly ensure their update and use them for the planning of their resources. Planners can also be seconded to the functions / departments to help with the scheduling and maintain the consistency with the Integrated Project Schedule.

Interfaces between lower level schedules are guaranteed through the identification of interface milestones that can be found in several schedules including the Integrated Project Schedule. Key interface deliverables such as

Requisitions (between engineering and procurement), delivery dates / Required-On-Site dates (between procurement and construction), need to appear clearly in the Functional Schedules.

Maintaining Consistency in the Schedule Hierarchy

It is essential at all times to maintain consistency between the different schedules used in the Project. This works both ways:

- from the Integrated Project Schedule towards the lower-level detailed schedules when it comes to changes in Project strategy, scope or execution strategy, and the influence of other functions or of the progress of other related activities;
- from the lower-level detailed schedules towards the Integrated Project Schedule when it comes to updating based on the actual progress of activities, updated forecasts or changes in execution sequences.

Consistency at all times is important and needs to be ensured by the scheduling team. It should be one of their most important points of accountability. This requires a tight communication between the scheduling team and the functions and can be ensured by organizing daily or weekly regular meetings.

When Project Managers Use Different Schedules for Different Purposes

This is one of the dirty secrets of Project management. While theory and common sense calls for a single, consistent set of schedules, for management, political or commercial purposes Projects sometimes (actually, quite often) have to maintain different sets of schedules – or, worse, voluntarily leave inaccurate information in the main schedules.

Most schedule handbooks avoid this question, but it needs to be addressed: Project Managers need to know why it happens and what can be done and what can't be done.

The risk is to lose control over the Project by piling up inaccuracies or by using voluntarily skewed schedules for decision-making.

The following broad types of schedule 'adaptations' can be observed:

- The operational-related schedules are generally made challenging hence optimistic,
- The budget-related schedules are generally conservative i.e. pessimistic,
- Contractual-related tweaks might be inserted in schedules (depending on the contract type and the party).

The main effects that can be observed as being at the source of these different variations are:

- From the management perspective, specific functions such as engineering, construction, operations, etc. tend to challenge their teams with tighter schedules (or, seen in another way, use the 'P10' best-case schedules and durations). This is a very natural tendency because functional managers need to have ready resources and equipment for the next phase to avoid any stand-by (hence need a 'soonest' date) and generally, want to challenge their teams and avoid any complacency regarding the duration of activities,
- Management will generally show very voluntary mobilization and demobilization plans to challenge the teams, and to sometimes fit contractual requirements which might not be realistic,
- More generally, the Contractor's or Owner's internal schedule for specific resources or equipment shared between Projects might not fit the required dates for the particular Project. Depending on the situation, because of some stakeholders the Project may not want to show this constraint; or on the contrary it might be in its interest to show clear overlaps so as to foster a resolution of this availability conflict between Projects,
- Because of contractual strategies aiming at gaining advantages under the contract, contractors or

clients may show specific sequence of activities or dates in the schedules that they know are not exactly accurate with respect to their execution strategy and capabilities (the effect will depend on the contractual details and whether it is a lump sum or reimbursable contract, as well as on the window mechanisms for the mobilization of major equipment), or for Project completion,

- Other stakeholder requirement or constraints might dictate the production of different schedules which can be more or less optimistic.

The fact that several schedules might be used by different parts of the Project and for different purposes is not an issue by itself as long as the Integrated Project Schedule effectively reflects the expected unfolding of events. In any case it is not recommended to entertain parallel and conflicting Integrated Project Schedules in particular for long spans of time.

Best Practice for Multiple Schedules

The best practice for using multiple schedules is the following:

- Lower-level schedules can differ from the Integrated Project Schedule as long as differences are acknowledged at management level and the main interface points remain consistent (for challenging schedules a buffer should be introduced at the end that is owned by the Project Manager and which allows to tie it up with the Integrated Project Schedule),
- There can be variations from the reference Integrated Project Schedule produced for the sake of detailed scenario analysis, but they should be clearly considered to be non-applicable versions even if they can be used as the basis of discussions with various stakeholders. They should be clearly marked as scenario analysis schedules not applicable for Project execution,
- There should be only one reference Integrated Project Schedule, mostly in line with how the Project is expected to unfold; only small tweaks can be

acceptable for tactical purposes as long as they are known and acknowledged. The Project financial forecast needs to be consistent with the Integrated Project Schedule at all times, or any discrepancy clearly identified and tracked (for example as an allowance),

- Different schedules can be produced for different stakeholders depending on needs (including scenario plans) but should only be produced at a higher level (200-400 activities). It would be extremely cumbersome and resource-intensive to update in parallel a second 'ghost' Integrated Project Schedule and that should be avoided as much as possible. Still, there might be situations where that could be unavoidable but it should be avoided by maintaining a proper connection and trust relationship between Contractor and Owner/ Client representatives.

Conclusion

From the onset of the Project, it is essential to have a clear and unambiguous schedule hierarchy that will be used throughout the Project. Project Value Delivery's recommended hierarchy will suit all Large, Complex Projects.

- The Convergence Plan has shown a proven effectiveness in many Project-driven organizations as a high level Project management and communication tool, in particular for complex situations. It is easy to implement and extremely effective,
- The Integrated Project Schedule is the centrepiece of the schedule hierarchy. It needs to be limited in size, well linked with a focus on the interfaces between functions, and well updated and re-forecast,
- Maintaining a Simplified Project Schedule in parallel to the Integrated Project Schedule is an asset for the Project that should not be underestimated,
- Detailed functional schedules are underlying detailed schedules that are used to program and monitor the daily work of Project contributors.

It is also important to remain in control of the possible variances between the different schedules due to their different uses and stakeholders involved, and make sure the Project Management Team knows at all times what is the expected course of events to be used as a reference: the Project's single Integrated Project Schedule.

What is absolutely essential here is to use the right level of detail for the particular purpose of the schedule. Don't fall in the common trap of believing that more detail is better and try to produce huge schedules that you won't be able to update or change! Force yourself to fit each schedule in a given level of detail to have the right tools for the job you need to do. Make sure, in particular, that the Integrated Project Schedule is balanced effectively between functions and does not typically exceed 2,000 to 2,500 activities.

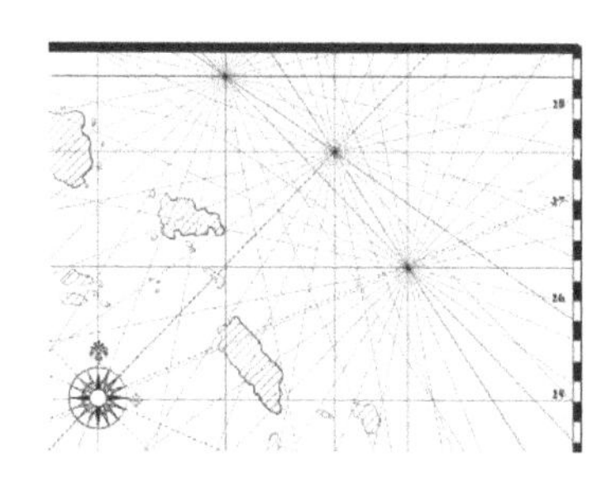

Chapter 3: Producing and Using the Convergence Plan

Chapter Key Points:

- Convergence Planning is a key strategic scheduling tool for complex Projects that is particularly useful to the Project Management Team.
- It is also very useful in terms of internal communication and alignment of the Project team. It is a useful vector for visual management.
- The Convergence Plan gates have to identify those deliverables that are really critical to the Project.
- Discipline must be applied when using the Convergence Plan: dates of gates should not change, and deliverables are to be effectively fully completed to be validated.
- Convergence Planning is scalable and can be used on more limited sections on the scope for closer control, or to track specific recovery plans.

Important Note: *Convergence Planning is mainly applicable to Large and Complex Projects and not so much to simple Projects, where it does not necessarily add significant value.*

Preparing a Meaningful Convergence Plan

Step 1: Identify the Workstreams and Critical Convergence Points of the Project

Ideally this exercise should be done at the tender/ planning stage before approval to proceed so as to minimize the number of critical convergence points required for the Project.

Notwithstanding such early optimization, it is important that at the Project kick-off, the Project core team spends time together to identify and discuss those main workstreams and convergence points between workstreams. This will foster an excellent shared understanding of the drivers of the Project.

This is best done as a facilitated workshop once the core Project team has been assembled, as it will serve as well as a session where the team will co-create the actual Project execution strategy.

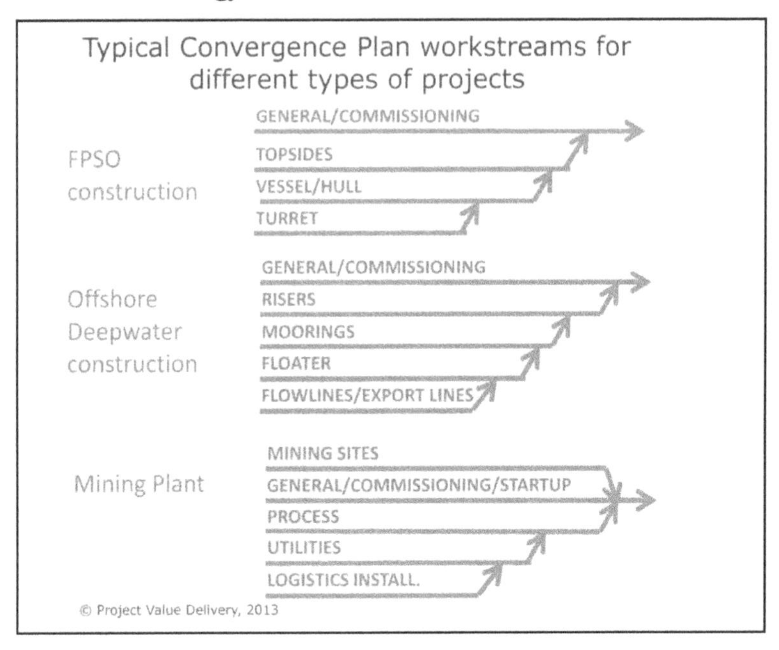

Figure 2: Typical Convergence Plan workstreams for different types of projects

Some key recommendations for this identification process include:

- Identify the different workstreams on the Project – in addition to a 'general Project management' workstream, there could be 2 to 6 different workstreams depending on the complexity of the Project,
- Identify the main deliverables required for the Project – those in whose absence, the Project cannot proceed; and generally, those outputs of significant chains of activities. The keyword here is

deliverables: what need to be identified are actual physical products or documents that need to be available, preferably corresponding to decision points (e.g. decision on an engineering concept or an option for installation). At this stage, do not seek to link those critical deliverables logically or to put them on a timeline,
- Categorize those deliverables by workstream,
- Further brainstorm whether additional early key decisions and deliverables would be required that would need to be added to the Project picture.

Step 2: Refine the Convergence Gates and Deliverable Dates

Until now we have only discussed about identifying key deliverables, not their scheduling. At this stage we need to start looking into dates. Start with the deliverables: the important part is to protect convergence points against natural variation in the production of the different deliverables required. In other words, we want to avoid that one deliverable takes hostage all the others and impedes Project progress. The best way to do that is to take the 'late' date where the deliverable needs to be available and impose a "buffer" between the deliverable availability and the actual convergence point. It is up to the Project team to impose a relevant buffer based on their experience of variation in the production of this particular deliverable.

Once all these deliverables are identified, as well as their duration and buffer, input all this information from the conventional Integrated Project Schedule. The critical convergence points can then be put on a timeline, along with the associated deliverables.

The following recommendations apply:

- While it is preferable that convergence gates correspond to actual physical convergence points, the deliverables do not necessarily need to be related except in terms of timeframe,
- Limit the number of critical convergence gates that will be tracked throughout the Project execution to ensure a real focus (rule of thumb is one convergence gate every 2-3 months for each

workstream); there should not be more than 4-6 main deliverables for each critical convergence point. If they are more, group them under a single set of deliverables,

- Convergence gates should be identified regularly throughout the Project; there should even be more in the early stages. In particular, there should be some at engineering stage and at procurement award stage because they are drivers for the Project final delivery.

Your Convergence Plan is now ready to be used as a key reference document for the entire Project team. Produce some attractive copies (adding Project-related images and illustrations) and hang them on the wall. They are great communication tools!

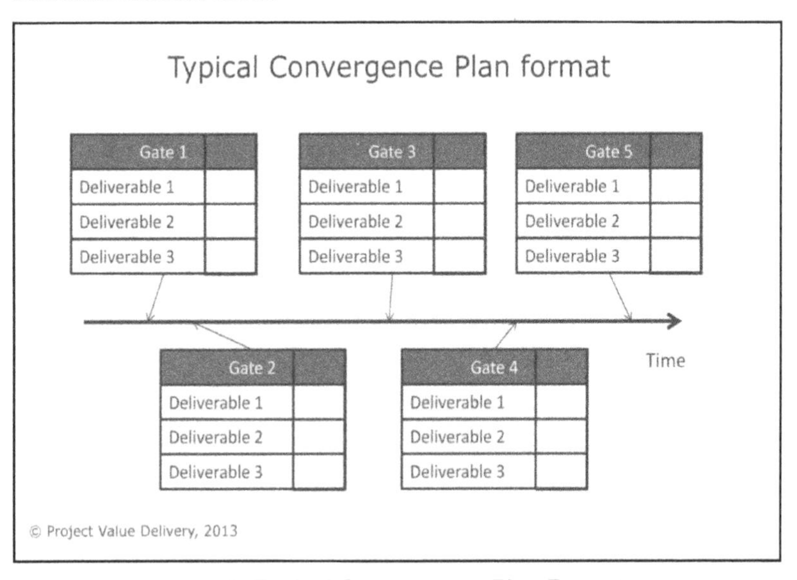

Figure 3: Typical Convergence Plan Format

Monitoring and Taking Action on Deviations

Update of the Convergence Plan needs to happen on a monthly basis. A coordinator needs to be nominated in the team (in general, the planner) to retrieve all the data on actual and forecast date of availability of the gate deliverables. A report on the status of those gates close to the current date is then included in the Project Periodic Report. An updated printout of the Convergence Plan needs also to be posted on the walls of the Project team's premises for internal communication purposes. It is a significant part of what is called 'visual Project management' and is an integral part of any Project war-room arrangement.

Figure 4: Real-Life example of a Convergence Plan posting on the project office walls (visual management)

Responsibilities

The responsibility for delivering the Project (or, the sub-Project) should remain with the Project Manager or the workstream delivery manager. The Convergence Plan should not introduce any doubt in that respect.

For each gate, a responsible person should be identified, the Gate Owner, who is responsible for monitoring and taking action to ensure that the gate is completed on time. Even if all the deliverables in the gate do not fall under his responsibility or budget, the Gate Owner is responsible for monitoring the Gate overall progress and taking action if it becomes obvious that a deliverable might be late. This will generally involve coordinating the work and enhancing communication across several functions. In addition, in most cases, the Gate Owners are the workstreams' Project managers and they have the authority to require project contributors to deliver.

For each deliverable, a person is then appointed to monitor that deliverable. For the sake of clarity, this does not include accountability as to the delivery of that deliverable. Still, this monitoring duty can be challenging because it will involve monitoring across functions and trades, and a heightened and wider duty of attention and diligence.

In those cases where the conventional schedule updating process might be slow or inaccurate, a more relevant and accurate monitoring is expected from the designated person, who liaises directly with the relevant sources of information. This independent monitoring will allow better quality of actual progress data. The sheer value of implementing that way an accurate monitoring of convergence plan's key deliverables is often considerable in Project organizations that have an inadequate schedule update process.

The responsible persons are mentioned on the convergence plan gate.

Typical Convergence Plan Gate format

Gate ID	GATE WS2-04		Gate Date	12/04/15
Gate Name	READY TO START UTILITY PACK #3			
Resp. Person	Mr. Package Manager		Gate Status	ORANGE
Key Deliverables		Required Completion	Resp.	Deliverable Status
Key Deliverable 1		12/04/15	Mr. A	GREEN
Key Deliverable 2		01/04/15	Mrs. B	ORANGE
Key Deliverable 3		20/03/15	Ms. C	RED
Key Deliverable 4		01/04/15	Mr. A	GREEN
Key Deliverable 5		20/02/15	Mr. D	GREEN
Key Deliverable 6		01/04/15	Ms. E	GREEN

Figure 5: Typical Convergence Plan Gate format

Monitor Regularly the Effective Convergence of all the Key Deliverables

At this stage, two keywords are important: discipline, and buffer monitoring.

Discipline

Contrary to the conventional schedule, the dates for the critical convergence points are set. They do not change with the actual progress. They can only change if there is such a significant modification of the execution strategy that it calls for a full rebaselining of the Project schedule. This will astonish newcomers to this method; still it is the only way to ensure that everybody takes these convergence points seriously. It is also one of the most effective ways to fight the 'virtual float' creation effect in conventional schedules, where when schedules move to the right, float is in effect created for all Project contributors which leads to diminished focus and sense of urgency (refer to Chapter 9).

Discipline needs also to be applied in reviewing the Convergence Plan progress regularly at the top Project leadership level, and take visible and massive action if needed.

Finally, discipline means that if a deliverable is 95% complete... it is not entirely complete and considered as such; only when a deliverable is really and demonstrably

100% complete can it be considered as done. This is because the last 5% progress is often long and difficult to achieve.

Float Monitoring

The key to convergence planning as an anticipation tool lies in buffer or float monitoring. Monitoring how the float on an activity compared to a fixed point (the convergence point) changes in time gives a great indication of whether the Project is effectively converging. By definition, an activity which float diminishes by one month every month will never happen!

This historical analysis of the float is not commonly practiced in conventional scheduling, and very rarely against a fixed date. In Project Value Delivery's experience, it is a key practice that offers a reliable early warning system for key deliverables that might not be completed when required.

It is also important at this stage to ensure a good quality of the float monitoring data: delivery forecasts should be reviewed and challenged to avoid responsible team members reporting wishes rather than reality.

The float monitoring techniques are expanded in Chapter 9 in the Float monitoring section.

Take Actual Action in Case of Deviation

Should the Project buffer monitoring system show that a deliverable is not converging appropriately, alarms should be raised and action taken early. Furthermore, if a critical convergence point is not going to be met, it should be an organization-wide wake-up call that goes all the way to the top of the organization. And this should happen even if it is a one week delay on a gate at the start of the Project!

It is then important to make sure the organization commits the right level of resources to rectify the situation and bring the Project back on track. It can mean a significant temporary commitment of resources. The extra cost will be more than compensated by the avoidance of a missed convergence point and the ensuing standby of most of the Project.

The discipline of respecting the critical convergence points is even more critical for the early convergence points. This may appear very unusual or even counter-intuitive for Project practitioners. What is the importance of a 2-week delay in the first convergence point, 3 month into a 2 year Project? Actually, it is really important because of the potential of the delay to snowball and should be treated as such. It should be remediated immediately to put the Project back on track and avoid complacency. Simply, successful organizations that lead complex Projects do that, like Toyota.

Application of Convergence Planning at Toyota

In the 1970s Toyota has revolutionized automotive project management, cutting substantially the time it takes to develop a new model. Toyota uses extensively Convergence Planning for its automotive Projects. This is the origin of the tool we describe here for Large and Complex Projects. Automotive projects are very substantial (several billion $) and complex, often involving the coordination of hundreds of suppliers.

Some best practices of project management at Toyota include:

Buffer usage: when designing the plan for the development of a new model, Toyota always introduces a buffer of 6 months between the date where is suppliers have to be ready and the start of production. This helps making sure all will actually be ready on time.

Contractual obligations on Gate deliverables: Gates' deliverables are made mandatory for suppliers and part of their contract.

Massive, quick and strong reaction to late deliverables: even at the start of a project and even for limited delays (e.g. one week), if a supplier misses a gate, Toyota will send over a significant task force to rectify immediately the situation. It knows that delays will snowball and that discipline is essential. This can involve sending physically a substantial team out of headquarters for a period at the suppliers' premises: massive, temporary resources are mobilized for rectification. This is complemented by a visible and very strong involvement of the company's leadership, up to the Toyota chairman even for lower rank suppliers.

Let's not be kidding ourselves – 80% of the effectiveness of the convergence method lies in effective, proportionate and timely action when a non-convergence is identified. The three first steps just provide the framework and a prioritized early warning system. The fourth step – action – will decide on whether the Project will be successful.

Scalability of the Convergence Plan Concept

The Convergence Plan is used successfully for full Projects, even very large and complex. It can also be used usefully for:

- Smaller sections of the scope that are critical to Project success and require a high level of coordination between contributors and/or functions (conversely, it is not very useful for low complexity situations, so that its application needs to be done with judgment),
- Recovery plans on Work Packages or sections of the scope that require high level of collaboration between contributors on a very constrained timeline.

For smaller scopes that remain of a high complexity, the benefits are similar:

- Allows for quick planning of the key activities that are the drivers of success and delivery (more quickly than establishing a full schedule),
- Allows sharing and creating adequate conversations between the different contributors about what is really important for the Project,
- Allows communication to the team on the key deliverables and their importance and enhances monitoring of what is actually happening.

Using Convergence Planning on sections of the scope that appear challenging due to their complexity is thus a good practice.

Summary: the Discipline of Convergence Management is the Key

When you are leading a Complex Project where many different interdependent contributors need to deliver in a coordinated manner, focus on convergence and be disciplined about it. Identify problems early by monitoring your buffers for convergence deliverables. Be relentless in ensuring that temporary additional resources get thrown in early if needed.

The Convergence Plan as a tool gives a high level view of the Project that can be easily communicated to the Project team, and allows the Project Manager to focus on those key high value deliverables which timeliness is essential to the success of the Project.

The most difficult part of its effective application is to have the discipline not to change the dates of the Convergence Plan gates once the Project is under way (unless there is a significant schedule rebaseline due to a change of scope or a major change in a Project milestone).

The Convergence Plan does not replace the Integrated Project Schedule which still needs to be designed, updated and forecast. Data from the Integrated Project Schedule will be used to monitor Convergence Plan deliverables.

Properly applied, the Convergence Plan provides the Project Manager with an uncluttered view of Project progress and enables the function to watch progress relative to the initial intent. It is also an outstanding communication tool within the Project team in terms of visual management.

Chapter 4: Producing the Integrated Project Schedule

Chapter Key Points:

- The Integrated Project Schedule is a tool for the Project team to support the execution of the Project. It needs to be aligned with the Project execution strategy and the Project decision-making process including the float management strategy.
- The Feasibility Study or Tender schedule is generally not suitable for Project execution and needs to be reviewed in depth at the start of the Project.
- The Integrated Project Schedule's detail needs to be limited. It needs to be balanced between functions (E-P-C) and properly inter-linked.
- When properly coded, an Integrated Project Schedule can produce the various views that will be useful to the Project team members.
- The Integrated Project Schedule also needs to respond to certain requirements to allow control of the Project: payment milestones, resourcing, etc.
- A schedule re-baseline is a massive exercise that goes beyond only schedule and needs to be apprehended as a new schedule development.

Setting up the Map of the Journey

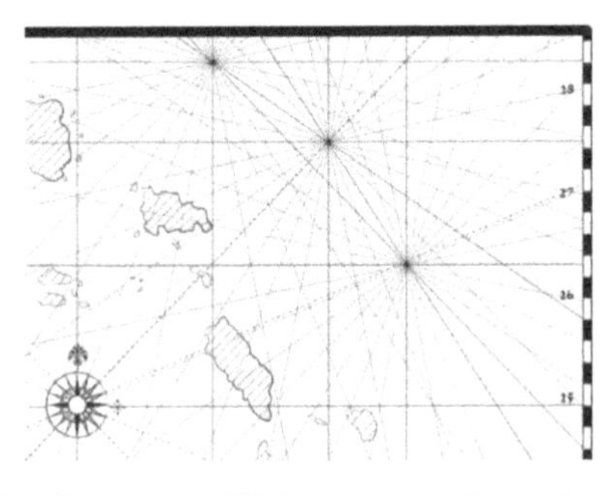

No-one can reach his destination without a map. Before sailing away it is necessary to make sure that we are purveyed with the appropriate set of maps to support our navigation. What we then require is a full set of maps of different scales depending on the phase of the journey. This set needs to be updated, reliable and consistent.

The schedule is the map of the Project. It is a primary monitoring and decision-making tool for the Project Manager and he should spend considerable effort and time to make sure its quality fits the required level – and that it is aligned with the strategy chosen for Project execution. In particular at the start of a Project, the Project Manager should spend a lot of time directing and reviewing the establishment of the Project schedule which is also a key component of the Project execution plan. We observe frequently situations in which the Project Manager has not spent enough time on the schedule at Project start-up (often because of being busy on stakeholder management issues, or purely due to competency issues) and this nearly always leads to Project failures.

Project management teams must understand that spending time to develop a meaningful Integrated Project Schedule is an essential investment which will repay itself multiple times over the course of the Project. We will not cover here the technical details of schedule production (such as the breakdown coding structure or the custom coding of the activities), but the cognitive work of ensuring that the resulting schedule is a logically-linked, appropriate description of the expected Project.

Who is the Integrated Project Schedule for?

This might seem an obvious question but it is not always clear in Project-driven organizations, because of the different expectations of various stakeholders.

It needs to be unequivocal that the Integrated Project Schedule shall first and foremost be a tool for the Project team.

Hence, while of course consideration should be given to the needs of various stakeholders, these needs should only be addressed inasmuch as it does not create significant burden on the primary purpose of the Integrated Project Schedule.

We have discussed the issue of how to deal with these pressures and the discrepancies between the different Project schedules in Chapter 2.

Producing the Integrated Project Schedule

Whether a detailed schedule has already been developed at tender/ feasibility stage or not, the Integrated Project Schedule needs to be deeply reworked or developed at the start of the Project to fit with the needs of Project execution and the Project execution strategy that has been chosen. In all cases it is a significant endeavour which will take a few weeks on Large Projects. How to drive the Project during these first weeks is explained in Chapter 5.

Schedule Principles - Why Cognitive Work is Required

There are a number of principles that need to be followed when producing the Integrated Project Schedule. Remember that we expect to have a ~2,000 activities schedule that describes properly the sequences of Project execution so that Project decisions can be based on a sound basis.

As soon as a Project exceeds a certain size, the Integrated Project Schedule cannot just be the juxtaposition of the Detailed Functional Schedules. Hence, choices must be made of what activities will be kept in the Integrated Project Schedule; what sequences of detailed activities need to be replaced with a representative higher level set of activities; and generally, how everything needs to be linked.

Choices are tough cognitive (and sometimes emotional) decisions. There is always the fear of missing something. The members of a function will always find good reasons why all their detailed activities should fit into the Integrated Project Schedule. But that would not work. Choices need to be made and they might feel tough. In addition they require a lot of thinking to make sure that what is shown in the Integrated Project Schedule effectively represents what is driving that part of the Project activities.

To make things easier, we present here some guidelines.

Underlying assumptions of Integrated Project Schedules

Project Schedules in general assume the following assumptions:

- *Resources that are allocated to the project have a normal productivity (possibly with a factor to account for location)*
- *There is no conflict between activities for resource allocation, or with other Projects.*

Review these assumptions and include the adequate correction factors if you believe they are not adequate.

The Integrated Project Schedule must be aligned with the Project's goals, decision-making process and execution strategy

When producing the Integrated Project Schedule it is essential to keep in mind the overall purpose of the Project and constantly refer to this high level picture when making choices about what should be captured and what should not.

A good practice in Project management is to develop Project goals, which are aligned with the stakeholder's expectations. They should be developed prior to the final establishment of the Integrated Project Schedule to ensure consistency.

The Project execution strategy and decision-making process (and the Main Contract management strategy) needs also to be reflected in the schedule. As examples:

- If the Project Strategy is to save a significant amount on Procurement, special attention shall be paid to the early steps of procurement (qualification of new suppliers, additional negotiation rounds, iteration with engineering),
- If a major uncertainty exists on the qualification of a particular equipment, specific early decision points need to be integrated in the Integrated Project Schedule,
- If cash flow is an issue the payment milestones need to be identified and brought as early as possible,
- If mobilization of equipment depends on some notifications or events, they need to be logically linked according to the expected decision-making flow logic,
- In general, links between activities should reflect the decision-making process and not just pure logic.

One major guidance point that needs to be given by the Project Manager is what approach should be followed for the management of the float. Should the Integrated Project Schedule rather represent challenging durations with a final buffer, or should it be built rather with comfortable durations and no centralized buffer? The decision will depend on commercial considerations and contractual issues (a buffer shown in a schedule might fall prey to the Client), whether the Project is lump-sum or reimbursable, and what approach the Project Manager wants to take with his team and other stakeholders.

General guidelines for Integrated Project Schedule activities

Here are some useful general guidelines for developing an Integrated Project Schedule:

- **Concentrate on the interfaces between functions:** the Detailed Functional Schedules do cover in detail the activities and linkages internal to the functions. Conversely, the Integrated Project Schedule should focus largely on the interfaces and links between functions, because it is the main tool where they will be identified,
- **The typical duration of an activity should cover a number of reporting periods:** so as to be able to do meaningful update and forecasting, activities should typically span over a number of reporting periods for the Integrated Project Schedule. Hence, detailed activities which are often short will not find their place in the Integrated Project Schedule. For large Projects, reporting is generally done on a monthly basis – meaningful activities should thus typically span over a few months,
- **Very early activities (in the first 2-3 months) and close-out activities should be represented by sets of milestones** rather than activities that would be tracked with physical progress,
- **The activities should be easy to update with actual physical progress measurement:** it is important to make sure that the activities in the Integrated Project Schedule correspond to work that will be easy to update and re-forecast – either because there is an underlying system providing progress figure (such as for example, a Master Document Register that computes actual progress), or reports are expected from suppliers and contractors that will provide the physical progress figure,
- **Vary the detail of the sequences of activity according to their criticality:** the level of detail used throughout the Integrated Project Schedule does not need to be consistent. Low criticality sequences of activities can typically be represented with a fairly low level of detail, while more critical

activities (because of their schedule criticality or because of a high uncertainty in their delivery) can be significantly more detailed,

- **Use custom fields to code the activities according to all the different views and filters you intend to use.** This is very important to ensure that the Integrated Project Schedule will be used to the maximum. A brainstorm with stakeholders and internal team managers is recommended to ensure a proper data structure. As a minimum it should be possible to sort/filter by function, facility system, facility geographical zone, main equipment, responsible office (if a multi-office Project team).

Even if the tender schedule looks very detailed it will almost certainly not be suitable to control Project execution (due to the way it is built and broken down) and needs to be deeply reviewed to suit the requirements of an Integrated Project Schedule. A significant work is still to be planned in that respect at Project start-up. Part of it will be to reconcile the schedule with the actual final scope agreed.
Do not ever take the estimating schedule as the baseline schedule for controlling Project execution without a deep review!

Putting Together an Integrated Project Schedule from Scratch

The best way, but the least common, is of course to start from scratch. For this, we recommend a top-down approach. A bottom-up approach (from the functions up to an overall image of the Project plan) will necessarily be more cumbersome, and less optimized towards the Project's objectives.

History of Project management has definitely demonstrated the superiority of the top-down approach to Project execution planning where all functions sit together to determine the optimal trajectory, versus the traditional bottom-up approach where each function states its best schedule and it is then aggregated. It is for example, typically the type of approaches that allows industry challengers to again and again turn around similar Projects with a much lower cycle time (the example of Toyota Project management practices and how it changed forever the face of the automotive industry is famous).

A recommended practice when developing a Project Schedule from scratch is thus first to have a good high level view of the Project sequence. This is achieved by brainstorming a Convergence Plan first. With a representative set of knowledgeable people from all relevant functions, such a Convergence Plan is easily produced in a day. This will get the Project management team thinking about what really drives the Project and give an overall reference framework.

Then, moving backwards and constantly asking the question "*what's really important/critical for the delivery of that piece of the Project?*" the most important sequences of events will appear readily. They can then be brought together consistently in a schedule.

Finally, the realism of the sequences can be ascertained at the detailed functional level. However, care must be exercised to ensure that the main drivers of the Integrated Project Schedule are effectively transferred at the functional level, otherwise the intent will be lost. In Project-driven organizations that have very strong functions (often called "weak matrix organizations" which are mainly function-driven with relatively weak Projects), there is unfortunately a high risk at that stage that functions deliver a *'self-centered'* work plan that does not fit the higher Project needs (that will be, for example, driven by resource utilization constraints and not necessarily the general interest of the Project). Proper communication needs to be ensured as well as sufficient authority of the Project Manager. A Best Practice in this respect is to mobilize an integrated Project office where all Project contributors are concentrated within reach of the Project Manager irrespective of their functional origin.

Upgrading an Existing Preliminary Integrated Project Schedule

It is more usual for a Project Manager at the onset of Project execution to inherit some schedule that will have been produced at an earlier stage – feasibility study or tender.

Because of a lack of time in the hasty days of Project setup and start-up, Project managers often do not take the time to question that schedule and use it straight away as the Integrated Project Schedule. It is a mistake, because the schedule that was developed during the feasibility studies or tender phases responded to very different objectives, and will generally not be suitable for Project execution. In particular, because it supports a pricing exercise, it will generally be too detailed and itemized. It might also not focus enough on the interfaces between functions that are important for Project execution. Logical links will also almost certainly not reflect the logic of the expected project execution's decision-making process.

A significant amount of work is thus needed to produce an Integrated Project Schedule that fits the needs of Project execution, even if there is a schedule available. The most common problem will be to remove details, because the schedule produced from a feasibility study or a tender will most probably be too detailed (detailed Functional Schedules are often not yet produced so that full detail tends to be borne in a single available schedule).

Once again, we recommend starting with establishing the big picture – the Convergence Plan. It will allow the Project team to have a high level strategic discussion on how the Project should be delivered, that will inform the later Integrated Project Schedule simplification and areas of focus.

Thereafter comes the most difficult of all: simplifying the Project schedule. Resistance will come from all fronts – not to forget the planner's resistance to remove details in a schedule that had to be amorously nurtured during months. It requires a lot of explanation, drive and time from the Project Manager to do this exercise properly, which therefore too often does not happen.

Main Issues in Large Projects' Scheduling: Detail Level and Balance between Functions

The Scourge of Excessive Detail

Throughout our consulting assignments on Large, Complex Projects in execution phase we've encountered too often an astonishing phenomenon: excessively detailed Integrated Project Schedules, impeding proper update, control and decision-making in the Project.

Planning must be done properly so as to enable to identify Project drivers. At the same time it must keep agility and reactivity intact as unexpected events will always happen during Project execution.

It is extremely important to optimize the detail level for Project planning between two extremes which are equally damaging for Project success.

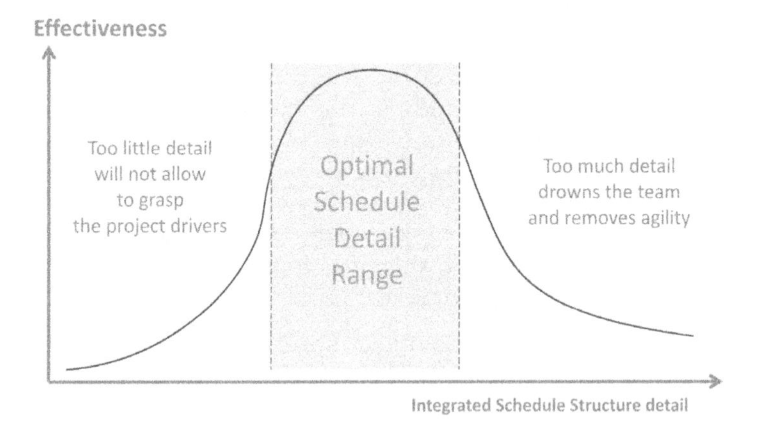

Figure 6: Optimizing the Level of Detail of the Integrated Project Schedule

Usual issues with too little detail in Project planning

Too little detail in Project planning will not allow the Project team to understand what the drivers of the Project really are. It will lead to the following issues:

- No proper identification of Critical Path and, more generally, of the drivers and constraints of the Project,
- No proper identification of resource requirements;
- Poor anticipation of interfaces between activities and contributors,
- Poor progress update due to insufficient granularity of baseline planning,
- Generally, excessive optimism regarding Project execution,
- Insufficient baseline for actual progress measurement and control of the Project.

The curse of too much detailed planning

More often we encounter excessively detailed planning in large, complex Projects. This results in overall Project execution plans often reaching 8,000 to 10,000 activities or more (and therefore 15,000 to 20,000 links), with printouts organized by function and not by deliverable, reaching the limit of the manageable.

This practice has a lot of drawbacks – for example upon closer examination, due to the sheer size of the schedule, independent reviews often show that the linkages between activities are not always representative of the logic of the Project resulting in incorrect logic. This is unfortunately often remediated by implementing an excessive number of constraints in the schedule network that would otherwise be superfluous, and will seriously impact the quality of the update and forecast of the schedule. Due to the large number of links, many can also be expected to be simply wrong.

As a consequence, the Project planning update is generally not representative of the actual activity progress that can be observed on the ground, and not any more used for decision-making by senior Project management! In addition the planner will spend more time trying to update and debug its schedule than trying to understand actual events and reforecast accordingly. This is of course a sure recipe for disaster – and it is astounding how often we encounter this situation during our consulting assignments!

Planning reloaded: what is the purpose of planning during Project execution?

Beyond the initial input into resourcing and budgeting, the Project plan during Project execution is ultimately used to take decisions. To achieve this, it needs to:

- represent effectively, at any time, the execution plan of the Project, taking into account any decision regarding changes of plans, new or changed activities or logic,
- and be rigorously and candidly updated as to the actual progress of the Project tasks.

A Project is a time-sensitive endeavour, and changes, updating and re-forecasting need to be done in real time within each reporting period to ensure that decision making is based on a sound reference. The scope of work to be done during each reporting period needs thus to be commensurate with the resources available to carry them out.

Just for the schedule update, if we suppose conservatively that only 20% of the Project activities are live at any given moment, and that to carry out a reliable progress update verified through discussions with the responsible people verifying the logic links etc., a planner can update 5 activities (and related links) for every hour, 8 hours a day for 10 days a month (the rest of the time being spent reporting and on other tasks), the schedule cannot have more than 2,000 activities in total. Depending on the planning resources and their tasks, there is thus a practical limit to what can be effectively managed in terms of Integrated Project Planning.

The agility imperative

In Projects in general, and even more so in complex Projects, planning agility is an imperative. It is necessary to be able to run alternate scenarios and Schedule Statistical Analysis (Chapter 7) as decision-making input on a regular and short notice basis. It is necessary to be able to modify the Project schedule as circumstances require it without spending 3 months for a schedule re-baseline because of the sheer work involved in re-linking all the activities. In general, the Project execution schedule needs to remain manageable to allow planning agility. It is another reason why at execution stage, the Project schedule complication needs to be limited. This issue is treated in more details in Chapter 10.

Summary about the Integrated Project Schedule detail level

While the issues related to too little detail are well known, it is important to underline that excessive detail is also a major impediment to Project success, because it drowns the Project team into lots of activity that does not add value when it comes to driving the Project execution:

- Project Team spends too much time planning and defers important long lead activities,
- Lack of plan overview and understanding of the main drivers due to excessive detail and possibly difficult to manipulate plan format,
- Excessively detailed schedule proves difficult to accurately update regarding progress at the expected update frequency,
- Lack of agility and reactivity when unexpected events happen.

Hence contrary to common knowledge, more detail is not necessarily good. A balance needs to be sought to ensure the maximum effectiveness for the team.

We thus recommend to aim around 2,000 activities in the Integrated Project Schedule. This number is for actual Project activities, this does not count milestones which can be added at will for purposes such as payment timing, interfaces etc.

Excessive simplification is the enemy of success. So is excessive complication. Strike the right balance. It is something of an art. Take the time to get the right balance at the beginning of Project execution. It is necessary for the success of your Project.

The Scourge of Unbalanced Schedules

We also encounter very often Integrated Project Schedules that are very unbalanced between trades/ functions. It generally reflects either:

- The history of the organization and/or its traditional focus (for example, certain organizations are engineering-driven and will tend to have much larger share of engineering activities; some others are traditionally more construction driven and that will reflect as well in the level of detail of that part of the Project);
- The comfort zone of the planner who will tend to spend time and detail excessively that section of the schedule that is the most familiar to him/her.

Such an imbalance is a problem. Because some functions are excessively detailed, while some others are scarcely represented, the linkages between functions will be incorrect and inconsistent, and the flow of activities required for the production of a particular piece of scope will be inadequate for piloting purposes.

A good Integrated Project Schedule is balanced between the different functions to guarantee the consistency of the linkages.

Float Management Philosophies

At the outset of the development of the Integrated Project Schedule it is essential to understand the float management philosophy that will be used. There are different ways of building a schedule.

- A first method considers realistic or slightly padded activity durations and produces a schedule network without any particular additional buffer,
- A second method is to consider challenging activity durations (best case), build a network accordingly and add a buffer at the end of the schedule that is owned and managed by the Project Manager. This buffer can be explicit or implicit (the latter being generally the case to avoid the Client to expect to manage it).

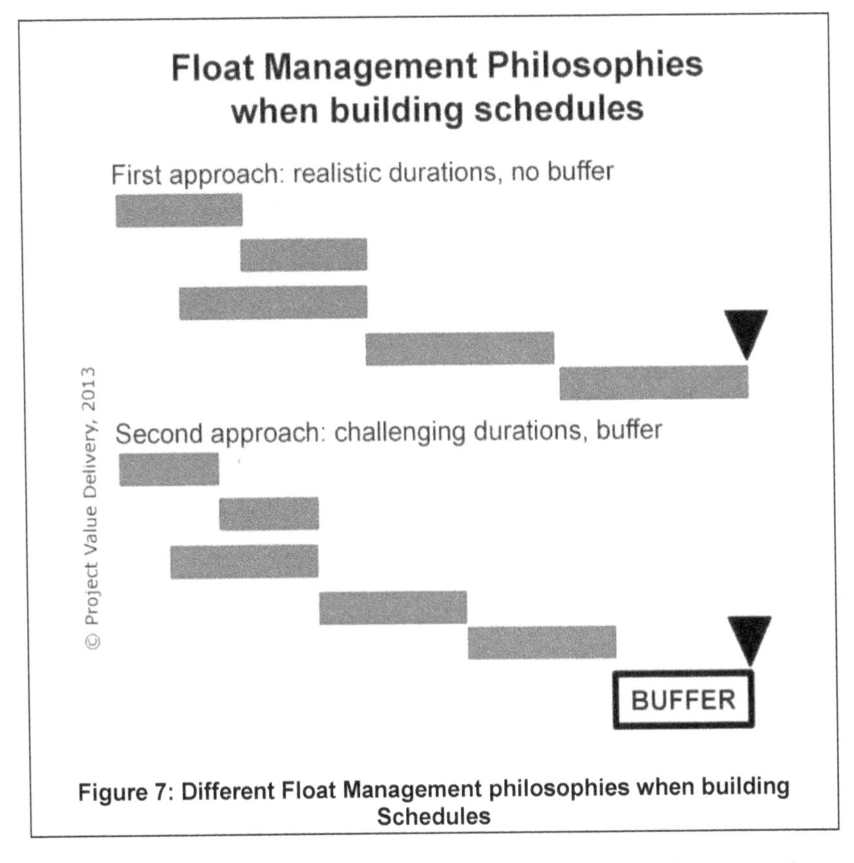

Figure 7: Different Float Management philosophies when building Schedules

The first approach is generally used in most Integrated Project Schedules, while the second approach is used in most detailed construction schedules (with the aim to make sure that the relevant logistics are ready early to avoid any standby on the construction organization and equipment).

There is considerable debate between Project Managers as to what approach should be used, for example if some intermediate approach such as using 'P30' durations and some buffer would be better. Some Project Managers advocate such an approach in the Integrated Project Schedule with the bulk of the buffer immediately prior to construction, which has also the advantage to challenge the contributors on early dates and protect the costliest phase of the project. Others underline the need to place buffers immediately prior to key Project convergence points.

Whatever the orientation that is chosen to build the Project schedules:

- The activity duration and float management philosophy needs to be consistent within a single schedule (although it can be different between the Integrated Project Schedule and lower level, more detailed schedules)
- The buffer/ float location is to be clearly known and identified even if some measures are taken to keep it hidden from the stakeholders if needed.

Specific details to be Included in the Integrated Project Schedule

Milestones to be Added for Various Purposes

Once the Integrated Project Schedule has been developed, it is adequate to include and link relevant milestones that are important for the Project Management team. These milestones can be filtered out of more operational schedule print-outs, still the information will be available and automatically updated. They do not count toward the 2,000 activities target.

Important milestones to add to the schedule include:

- Invoicing milestones to the Client as defined by the contract (the download of their updated dates from the schedule will feed the cash flow forecast),
- Key contract milestones, in particular, Client Provided Items availability dates, mobilization window dates, and dates where Liquidated Damages apply,
- Convergence plan deliverables availability dates, to be able to download this information into the Convergence Plan and do float analysis (ref. Chapter 10),
- Convergence plan gates dates, and in general, key Project convergence points, so as to be able to observe any change and react in advance. Those specific milestones need to be included as well in the Simplified Schedule. On this matter, it is important to focus on early convergence points to increase the

odds of Project success; whereas later milestones are often linked to contractual obligations, but have less influence on actual Project execution success.

Weighting of Activities

To derive relevant progress figures, activities in the Integrated Project Schedule need to be weighted. The most relevant way to weight them would normally be to use their estimated cost. On lump sum Projects this is an issue in particular if the schedule source file needs to be sent regularly to the Client. The Project Control Manager needs to generate a list of dummy activity weights based on the price (and on the value of the different milestones, which are generally associated with specific parts of the work). When weights are derived from price milestones, the schedule file can safely be sent to the client. There will be a discrepancy with a pure cost-based weighting but they will remain limited compared to the overall picture.

It is not too much an issue to leave up to 20-25% of activities un-weighted when doing this exercise as long as enough activities have been weighted in each function and each scope package to generate adequate S-curves.

Weights are generally not updated during the course of the Project even if some element of the budget would show a very significant variance, as it would cause inconsistencies in the progress measurements, and because the impact of such change would anyway be small with regard to the entire Project's progress, unless there is so much change that a full rebaseline is required.

Proper Schedule Coding

Proper schedule coding is necessary for two purposes:

- Allowing to use different views and filters of the schedule to give different contributors and stakeholders the view they need to take decisions (refer to Appendix 5 for discussion on the WBS dimensions to be included),
- Enabling Earned Value Analysis on those areas of Project execution that are suitable for this type of analysis.

Don't underestimate the value of using different schedule views

Modern schedule software such as Primavera® use a data-base for data management, which allow for various views and filters that can cater for different uses.

Producing the right views and printouts of the Integrated Project Schedule for the different users on the Project has a value that is often under-estimated. Project managers and delivery managers require schedule views by subsets of scope showing typically the Engineering – Procurement – Construction - Commissioning chains of events. Functional managers require a different view by function (and possibly, sub-function) to organize the work and the appropriate resources. Others might only require to view a subset of the Work Breakdown Structure (refer to Appendix 5).

It is very easy to produce these different views from the main schedule. View templates can be pre-established which include specific filters, show certain information, and order the activities as required. Where most Projects only produce a single schedule print every month, others do produce as many printouts as required to optimize the value of the updated schedule database for all Project contributors.

Implementing Earned Value Analysis

Earned Value Analysis requires crossing physical progress data (from schedule updates) and cost data (from cost control). To achieve this, the same breakdown structure needs to be available in both schedule and cost control systems to allow this data reconciliation to happen easily.

This interface is not always used. We describe it here quickly for the sake of completion and because it can be useful, in some instances, for specific functions. To allow Earned Value management techniques to be deployed on the Project, the schedule should ideally be coded as per the final Cost Breakdown Structure (CBS). However a specific schedule Work Breakdown Structure will have been established at the feasibility or tender stage following a different logic and it is cumbersome to re-code the entire schedule according to the new CBS (in all cases, this re-coding should be done in a separate user-defined field to keep traceability with the original schedule). It is sometimes useful to take the time to add the cost CBS codes, for certain cost types only, on large Projects so as to benefit from the power of Earned Value management measurements.

The cost types where this coding is useful for most organizations include:

- Engineering activities,
- Fabrication activities,
- Construction activities,
- Commissioning activities.

Resourcing

Resourcing is an important component of schedule preparation in Large and Complex Projects. In particular, it allows checking that resources are sufficient in number to make the schedule realistic, or whether the work can be done if there is a particular constraint on the number of people that can work at any one time on the Project (such is often the case on remote sites).

If resources cannot be mobilized in sufficient number to execute the work, the process of *'resource levelling'* needs to be applied, which requires a lot of experience (resource levelling needs to be done manually avoiding the automatic routines of schedule software to understand what is happening). The schedule duration will of course increase in the process (if these issues apply to activities on the Critical Path) and/ or the available float diminish for part of the Project.

The activities that need to be resourced are those for which progress (Earned Value) is directly linked to resource availability and/or productivity. Hence, resourcing should be potentially considered for the following activities on most Projects (most of the time it should be a requirement for the relevant contractors' detailed schedules):

- Design Engineering activities, and detailed design in particular,
- Fabrication activities,

- Construction activities,
- Commissioning activities.

Schedule Rebaselining

We have purposely included the issue of schedule rebaselining in this Chapter about producing the Integrated Project Schedule, and not in the Chapter about updating. This is because schedule rebaselining should not be mixed up with a mere update. It is a serious matter that should be considered on the same level of importance than actual production of a schedule.

It is essential that rebaselining remains a rare event, because of the importance of having a stable reference throughout the Project execution with which to compare progress. As Dilbert says in one of its cartoons, a schedule that changes too often should not be called a schedule, but a calendar!

Schedule rebaselining should hence only be contemplated when there have been such serious changes in the Project execution context that make the current Integrated Project Schedule irrelevant in terms of logic. That should be rare and it is a very serious decision to be taken by the Project manager, because of the far reaching consequences. In any case, rebaselining should not just be motivated by delays compared to the baseline if these delays do not change the general Project execution logic.

When rebaselining, we recommend to first review the Convergence Plan outline to check whether the Project strategy should be amended at a high level. This should be done during a workshop involving representatives from all functions. Then from the result of this execution strategy discussion, the Integrated Project Schedule, and subsequently, the Simplified Project Schedule and the Detailed Functional Schedules can be amended as required.

Conclusion: Take the Time to Devise a Proper, Usable Integrated Project Schedule

Project managers and their teams have all sort of excuses about not having the time to lead a proper review and development of their Project's Integrated Project Schedule. It is true that at the onset of Project execution the amount of tasks to be done is daunting.

At the same time, would you sail for a long voyage without a proper map that would serve you to take navigation decisions?

Producing a useful Integrated Project Schedule is arguably the most rewarding investment that can be done at the start of Project execution. The investment will redeem itself through proper positioning and relevant decision-making throughout the Project life.

Take the time to devise a proper Integrated Project Schedule. Do it as a team with representatives from all functions, aiming at the Project purpose and goals. It is really worth it.

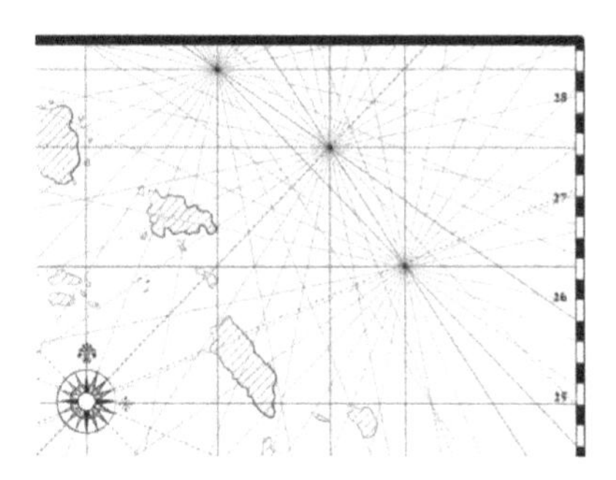

Chapter 5: Schedule and Project Lifecycle

Chapter Key Points:

- At Project start-up, develop the Convergence Plan as soon as possible and drive the schedule with deliverable dates. It will often take 2-3 months to establish the full-blown baseline Integrated Project Schedule, the associated detailed document register and detailed construction schedule.
- At all stages it is essential to keep consistency between schedule, cost time-phasing and Opportunities & Risks.
- At close out stage, drive the schedule with deliverable dates and punch lists.

In this Chapter we will focus on how the different components of the schedule hierarchy are developed and should evolve throughout the Project lifecycle, from early tendering to Project close-out. A check-list in Appendix 3 summarizes the key aspects that need to be focused on by the different functions.

Feasibility Study/ Tendering Stage

The quality and the method of schedule development at feasibility study / front-end engineering / tendering stage depend a lot upon the available time. Small to medium size Projects tend to have short gestation times. Very large and complex Projects tend to have long feasibility phases or tenders until the actual start of execution. The focus and the approach will hence be different depending on the situation. In all cases, the schedule that will be developed for the feasibility study or tender – driven by an estimating purpose – will not be directly usable for Project execution control (ref. Chapter 4).

The tender or feasibility schedule will focus on the EPC critical network leading up to construction, and then the construction critical network, so as to ensure that the proposed completion dates can be achieved. It necessarily remains at high level with sufficient attention devoted to the presumed Critical Path. Estimates of resources required to complete the overall Project in time (in particular, for engineering) are then guessed based on a proposed staffing plan, benchmarks and experience.

For short, in simple projects the focus in developing the schedule is generally upon both:

- The Critical Path – related to the commitment to Project completion date; and
- The construction phase, because of the high cost impact of that phase of the Project, and the relative minimal importance of engineering and procurement.

For Large, Complex Projects, at high level, what drives the Project Critical Path for a particular execution strategy has to be identified at the feasibility stage. It can either be:

- Long lead items (such as large quantities of bulk material such as pipes or specific highly complicated equipment),
- Specific complex development and qualification processes, typically in the field of innovative process features, specific raw material processing issues, specific parameters related to construction material such as for example, high resistance concrete or high-technology welding,
- Availability of a key construction enabler which is very rare. The schedule must be built around its availability or minimize the number of mobilization/ demobilizations.

A tender schedule must include relevant float even if it is not shown to the Owner/ Client, so as to improve the possibilities of timely delivery for both parties, and diminish the risk of Liquidated Damages for the Contractor and loss of revenue for the Owner/ Client.

A tender schedule must also be aligned with the tender negotiation strategy. One aim of a Contractor is usually to demonstrate to the Client that the contract needs to be awarded soonest. Depending on the situation, specific chains of events can be highlighted to bear on the negotiation. Care must also be taken on the milestone payments as the tender schedule will be used as a basis for the planned dates.

A detailed construction schedule is often developed at tender stage because of the intrinsic high cost of this part of the Project, and possible constraints regarding resources availability or worksite congestion. The Construction schedule then drives the logistics schedule, which must fit the expected construction progress rate. The role of the Construction Manager is essential in the timings estimates that will be used as a basis. However, in Complex Projects it is also important to spend enough time on the Engineering and Procurement phase, which will often constrain the Construction phase.

Even when there is time to develop a quite comprehensive and detailed overall schedule for the Project, because it is primarily developed from an estimating perspective it will generally not contain the links, sequences and intricacies that need to be considered during Project execution. Refer to Chapter 4.

When tenders or feasibility studies have long durations there is a tendency to develop further the Project schedule up to a point which is sometimes excessive. This is sometimes to keep people occupied on the tender. Long tenders or feasibility studies should not necessarily mean excessive detail on schedule and costing, beyond a point that would be much more detailed than the intrinsic uncertainty of the exercise.

Project Start-Up and Engineering Stage

For the execution of large Projects it will take some time (2 to 3 months) to go through the motions of Project start-up until the full baseline of the Project is established (which includes not only the Project schedule but also more generally, the definition of an execution strategy (including how to exploit key Project opportunities and address key Project risks), the Project execution plan, the Master Document Register and the outline of a detailed construction schedule).

A common mistake of many tenders is to assume there is no ramp-up period during project start-up and assume that the project is able to progress at 100% resource loading at the very start. This is not often true, and this often causes delays even as the Project has not really started yet.

Some Owners/ Clients require the immediate production of a 90-day start-up schedule detailing all the activities that will be carried out in this start-up phase.

The First Months

In the first 2 or 3 months where the Project does not have a firmly established baseline Integrated Project Schedule. It is still possible to move the Project forward by:

- **Ensuring a proper handover from the proposal / feasibility study team.** This handover should be documented and an updated version of all reference documents used in particular for estimating should be made available to the Project.
- **Progressing the order of long lead items, support infrastructure** (camp, access road, airport runway...) **critical works** (e.g. earthworks) **and key base parameter studies** (e.g. geotechnical) already identified at pre-execution stage as quickly as possible. For very critical items the organization will often have invested in preparing this process in the few previous months' (engineering and procurement pre-award activities), leading to a position where the Project is ready to pass the Purchase Order immediately upon Main Contract award,

- **Developing the Convergence Plan in the first 2 weeks** or as soon as the Project management team has been mobilized, so as to align the team around the priorities of Project execution and guide the development of the Project execution strategy and baseline; and in addition, drive the first priority activities until such time the Integrated Project Schedule is established,
- **Focusing on achieving milestones and deliverable dates** as in that part of the Project, progress measurements and S-curves won't make sense and will not have been setup properly anyway.

The Integrated Project Schedule can be developed in parallel of the development of the Project execution plan, taking into account the contractual strategy that will be followed by the Project.

Project Baseline Establishment

The Project baseline is finally established. A baseline does not just cover schedule but all the different aspects of Project planning (execution planning, in general and for each function; organization; cost forecast and time-phasing, etc.).

The Project baseline for large Projects should be established latest 3 months after Project kick-off. The Convergence Plan dates need to be realigned with the baseline at that stage.

Engineering Stage

During the engineering stage, a particular focus needs to be exercised on the engineering deliverables that interface with other functions:

- Requisitions for procurement,
- Scopes of Work for Service Contracts (including logistics and on-site services),
- Key construction procedures – for the construction equipment and aids that need to be procured and for the interfaces with construction assets modifications (if any),

- Vendor data (feedback from Procurement into Engineering).

In effect, vendor data is sometimes a required input for the finalization of engineering. It needs to be generated by the Vendors shortly after the Procurement activities, and needs to be linked back into the Engineering process at a suitable time.

These are typically documents that need to be identified on the Integrated Project Schedule, and not necessarily the detail of the engineering work that leads to these documents. Their schedule should be driven by the procurement cycle and the required availability date, and engineering activities fitted around this requirement, rather than the contrary.

Project Procurement Stage

At the procurement stage the actual lead time of procured items need to be included in the schedule, in particular when there is little float. They will be confirmed in the final stages of negotiation with the suppliers (substantial buffers might need to be included when lead time is a substantial commercial factor). The full procurement cycle needs to be considered, from bidding, negotiation, up to delivery customs-cleared.

Activities leading to the start of manufacturing need to be driven using milestone dates as they cannot be related to direct productivity; while manufacturing activities can be tracked using productivity-driven tools such as S-curves and productivity ratios.

For complicated items procurement and for contracts for services, good quality detailed schedules need to be requested from the suppliers and contractors that need to be updated regularly. These schedules also need to be challenged by the expediters in charge when they are received, in view of evidence of actual progress that can be available otherwise.

It is important not to underestimate the lag between the issuance of a Request for Quotation and the actual placement of a Purchase Order; and subsequently between the Purchase Order and the actual start of manufacturing. These lags can be heavily influenced by Owner requirements to review documents or participate to activities, including possible approval by the Owner of certain documents prior to the start of production for complicated equipment. Similarly it is important to include sufficient durations for Factory Acceptance Test and shipping activities.

Particular areas that need to be reflected in the Integrated Project Schedule include:

- Supplier data for Engineering completion,
- Interfaces with other activities or function during the manufacturing itself: e.g. expected late final engineering; Owner/ Client interface,
- Expected delivery date and location (including time for logistics where appropriate).

Specific attention needs to be exercised on contracts for services in particular those that play a decisive role in the later productivity of fabrication or construction, in particular if extensive qualifications are needed.

A schedule driver might in this case be the required test material (e.g. steel test material for welding, mine representative ore for process, etc...) as there might be substantial lead time. This is in particular the case for specific materials (e.g. forgings, exotic material, unconventional geometries and thicknesses) as they need to be produced first as part of the larger order for base material.

At Project procurement stage it is essential that as the schedule is updated with the suppliers' and service contractors' data, the Project cost model be updated consistently.

Project Fabrication Stage

Fabrication (fabrication of modules or other items outside the construction worksite) is a specific phase of the Project which is often schedule-critical because:

- It is linked with the site construction on the back-end (hence, a very high sensitivity in case of any possibility of site stand-by, or even for sites that are very exposed to seasonal weather, the risk of the postponement of a construction campaign),
- Site Integration tests require the mobilization of specific equipment,
- It often depends on the other end on the actual delivery of the required material and equipment (e.g. valves, forgings, etc.) as well of the related engineering (drawings),
- In some cases, when specific fabrications involve low total quantities and fabricators have concurrent activities for other clients, actual progress can be sensitive to the actual availability of resources for that particular fabrication.

The Integrated Project Schedule needs to reflect adequately the interfaces with engineering design, procurement of material and equipment, and construction.

Activities leading to the start of fabrication need to be driven using milestone dates as they cannot be related to direct productivity. Fabrication activities can be tracked using productivity-driven tools such as S-curves and productivity ratios.

The fabricators should be requested to produce and update a detailed fabrication schedule. This schedule needs to be resourced by categories of trade (e.g. welders, painters, electrical technicians etc.) so as to establish a productivity baseline and resource curves. This can then be used to measure actual productivity and infer sound forecast estimates for the actual delivery.

Issues that need to be watched for common delays include:

- *Specific welding, painting and coating qualifications which can sometimes take much longer than the actual fabrication,*
- *Exotic material fabrication will be constrained by the limited availability of 'clean' workshops and qualified personnel,*
- *Factory Acceptance and Site Integration Tests which take significant time and footprint.*

Project Construction Stage

During Project construction, a detailed schedule should be maintained by the Project Construction Manager. This schedule is generally a stretched 'best early case' so as to make sure that the most expensive construction assets do not incur any standby even if they are to operate at a maximum productivity. There will thus be temporarily some discrepancy with the less detailed Integrated Project Schedule, which is generally manageable at a macro level.

On the other hand, the Integrated Project Schedule needs to reflect the following interfaces:

- With construction engineering, in particular for the procurement or fabrication of relevant installation aids,
- With procurement and fabrication (or Owner's supply of free-issued material or equipment) for items to be installed,
- With service contracts for the finalization of qualification and the early mobilization and logistics for related equipment,
- With the logistics department activities for the organization of the flux or equipment, material and people to / from the construction worksite (included if required, organization of local accommodation).

Because of the inherent high daily cost of construction operations, proper time-phasing of the costs, consistently with the schedule, is essential during this stage.

It is also important to organize the generation of the as-built schedule prior to the start of construction operations. A detailed as-built operations schedule should be updated daily from the sites' Daily Progress Reports. The objectives of this as-built construction schedule include:

- *Benchmarking for future Projects,*
- *Performance analysis,*
- *Backup for invoices,*
- *Backup for claims and Change Orders.*

Project Close-Out Stage

Driving the schedule during the close-out phase requires punch lists and lists of key deliverables, associated with expected dates. These can be reflected as milestones in the Integrated Project Schedule and more detailed schedules. Progress measurements and S-curves have no utility at that stage.

Project close-out activities also include the generation of an as-built Integrated Project Schedule and detailed as-built operations schedule, which are important deliverables for the organization's benchmarking efforts.

In general the focus of the Project Manager should be on the proper close-out of the Project in all its aspects (documentation, non-conformances, suppliers' and contractors' accounts etc.) and this typically requires more time and effort than generally anticipated. There is also often a strong pull to other more exciting activities. The schedule forecast for the Project close-out needs to account for the actual anticipated duration of this phase. The schedule process must be a key support to the Project Manager to ensure that all activities are properly closed-out in time.

Relevant Schedule Reporting Tools

During the Project lifecycle, various reporting tools are being used.

The Convergence Plan is used throughout the lifecycle; it is important not to use it only as a reporting tool for the gates past, but also by forecasting the upcoming gates of the next few months so as to be able to anticipate and react if a key deliverable was suspect to be late.

S-curves are often used to examine the overall progress of the Project. They are only useful between 20 and 80% progress and meaningless beyond these boundaries. S-curves can be produced for sections of the scope. Important good practices include:

- Ensure to always keep visible the different baseline S-curves in particular if the Project has been rebaselined, as a reference,
- Most Projects show an 'early' and a 'late' S-curve (as defined by the Integrated Project Schedule network) and remain confident if they stay above the 'late' curve. It is a bit complacent to stay close to the 'late' curve as the amount of float is minimal. The slope of the current progress S-curve does give some indication of the trend, which is somewhat influenced by the progress measurement framework and is not always reliable for forecasting,
- It is possible to derive productivity ratios from physical progress-based S-curves for forecasting purposes (refer to Appendix 6 – Earned Schedule Management).

S-curves are general bulk progress measurements that do not describe in any manner whether the Project is being held up by a particular critical activity and in that sense, cannot be used as an early warning system.

Dated lists of documents, milestones and activities are used at the start and at the end of the Project (in those areas where S-curves are meaningless) as the most effective way to pilot the necessary activities. Punch lists are used at the end of the procurement, fabrication and construction activities.

Conclusion

Schedule-wise, the focus of the Project Manager changes in the different stages of Project execution.

The Integrated Project Schedule should always be focused on interfaces between Project functions (Engineering, Procurement, Construction and Commissioning) and entities such as Owner, suppliers and service contractors.

Detailed schedules become essential in particular for the manufacturing of complicated equipment, fabrication and construction operations. They help the crews on the ground deliver their scope, while the Integrated Project Schedule covers the interfaces with the rest of the Project.

Finally the Convergence Plan should be the very first schedule-related document to be established in the very first few days of Project execution to help the Project management team drive the Project all along. It allows adequate focus on the key success drivers of the particular Project.

Chapter 6: How to Easily Check the Quality of Execution Schedules

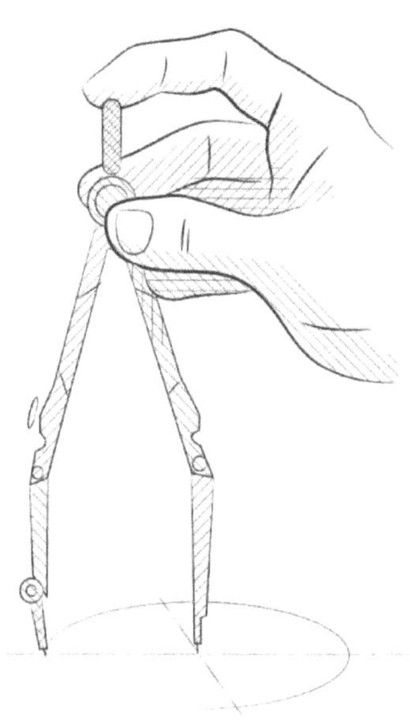

This Chapter intends to give to Project Managers quick and easy tools – the same we use at Project Value Delivery in our consulting assignments – to assess the quality of the schedules that have been produced by his/ her team.

Chapter Key Points:

- It is critical for a Project Manager to be certain that the tool used for navigation – the Integrated Project Schedule – is built properly. Otherwise, what can be the confidence in the current status and forecast?
- Schedule print-outs need to show systematically certain basic information: total finish float, baseline, and show a clear Critical Path.
- Further checks need to be carried out on the schedule to check that it is properly set: a minimum number of constraints, an adequate ratio of links to activities (between 1.5 and 2.2), a very low number of activities without successor or predecessor.
- Checks can also be carried out regarding the realism of the schedule, in particular concerning resources.

It is very easy for a Project Manager that has no previous scheduling experience to be blinded by the mysterious and powerful analytical scheduling tools, and the nice colours of the print-outs. He might forget to check that the schedule is actually consistent, appropriately linked and overall a sound basis for decision-making.

We have seen in particular many instances where Acrobat pdf or paper printouts of the schedules were used with confidence throughout the Project teams and with the clients and stakeholders, whereas a quick check on the native schedule showed a completely inconsistent, unlinked schedule that was not worth the paper it was printed on.

We strongly encourage Project Managers to invest the time to be familiar with scheduling tools. They are conceptually not complicated, and require mainly logic and application. A half-day demonstration by an experienced planner will be enough to understand the main mechanisms that are at play.

There are two issues that need to be checked when it comes to schedules: technical correctness and realism of activities' estimated durations.

***Important note:** the content of this Chapter is only strictly applicable to project execution schedules. Feasibility study or tender schedules may not be built following the same rules for the following reasons:*

- *Detail of activities unknown or assumed in some areas, replaced by placeholders*
- *Logic of the schedule might be different, e.g. schedule built from the delivery date backward to assess a required start date, utilization of positive or negative lags to avoid detailing, etc.*

However, upon start-up of execution it is essential that the schedule format and linkage be converted to a proper execution schedule, following the decision logic and with linkages moving toward the future, which is the only way to properly progress the execution schedule. Depending on the format and level of development of the feasibility/ tender schedule this may be a very significant task.

Schedule Correctness Checks

In this section we will be concerned by the technical correctness of the schedules. Some issues can be seen on print-outs whereas some others can only be observed on the native files.

Schedule Balance Between Functions

This issue has been already discussed in Chapter 4. The Integrated Project Schedule should be balanced between main functions (Engineering – Procurement – Construction – Commissioning). Whether or not it is the case is a very simple test for the maturity of the schedule. Very imbalanced schedules will certainly fail to qualify for proper Integrated Project Schedules.

What all Schedule Printouts Should Show

All Projects should have a standard way of presenting schedules. At Project Value Delivery, we recommend all schedules bar-chart (Gantt chart) printouts to include the three following information for each activity (in addition to the usual information such as activity ID, start date, finish date, duration and name):

- The Total Finish Float as a column,
- The baseline activity schedule as a line underneath the updated activity bar,
- A specific identification of the activity sequence defining the Critical Path.

We will now explain how this information can be used as good indicators of the schedule's technical quality.

Total Finish Float

The Total Finish Float is the total amount of time an activity can be delayed without impacting the Project finish date (it is different from the free float which only measures by how much an activity can be delayed without impacting the start of the successor activity). It is a good measure of the "slack" that the Project Manager has on this activity from the overall Project perspective.

Provided the schedule is not too constrained, an activity that is on the Critical Path should have a Total Finish Float of 0. A positive Total Finish Float indicates that there is some slack as to the start date of the activity and measures the sub-criticality of that particular activity or activity chain. A negative Total Finish Float indicates the presence of constraints, possibly regarding the Project completion date or other intermediate constraint.

Total Finish Floats are extremely useful because excessively high values (as a guideline, over 100 days on a multi-year Project) or awkward values will hint directly at lack of linkages within the schedule. If an activity has no successor the program will simply calculate the Total Finish Float as the difference between the activity finish date and the Project's end date!

It is thus very easy by looking at the Total Finish Float column in the printout to have quick check of the quality of the work of the Project planning team. A well-built schedule should show consistent values that should remain relatively low.

In addition, having this information systematically on a schedule printout for well-connected schedule networks is extremely useful to Project management because it gives a good idea of how subcritical chains of activities influence the Project, i.e. by how much they can be delayed without impacting the Project finish date. This information can directly help to drive Project decisions.

Note on Negative Float

It happens sometimes that schedules show negative floats between activities (total float or even free float). This can only happen if there is an artificial constraint in the schedule. A schedule that would be minimally constrained should not show those negative floats.

It is sometimes useful to constraint schedules with fixed milestones for very critical activities driven by stakeholders outside the Project (such as the mobilization of an expensive asset or the shutdown of an existing facility). This will allow to monitor the available float and a negative float will be an immediate warning requiring prompt rectification action, or contact with the relevant stakeholder.

In any case, negative float situations should be eliminated because obviously this cannot correspond to a realistic situation, unless there are some commercial reasons to show them.

Baseline vs actual or forecast

Comparing the baseline and the latest actual or forecast schedule of an activity is extremely valuable and it is amazing how few organizations do require this to be shown systematically on their schedule print-outs (maybe because of the unbearable discomfort of showing the slippages compared to the original plan!). It is an excellent visual cue of how the Project is unfolding compared to the baseline plan.

From the technical perspective, it is an excellent indicator of the quality of the schedule update. If too many activities stick to the baseline schedule, it probably means that the update and forecast processes do not work properly. If that is the case mainly for future activities it shows that the forecasting is not effective and that there might be some links missing, in particular if the current actual work deviates from the baseline at the same time.

Critical Path

While it should be a basic check, it is sometimes overseen. A proper schedule should of course have a Critical Path, i.e. the sequence of activities that adds up to the longest overall duration and thus drives the Project completion date.

The Project Manager's focus should of course normally be on those critical activities because any delay will impact the completion date (and any early finish will create opportunities).

Checking whether the Project schedule effectively identifies a Critical Path is another good check that the schedule is properly linked. Critical Path activities are generally identified by the colour red on a schedule print out.

Normally the Critical Path is strictly defined by activities with a total float of 0; however on large Projects it can be required to identify all activities with a total float of less than 15 or 30 days, which adds on some activities which are not strictly on the Critical Path and can somehow blur the picture. Anyway, a clear Critical Path in the form of a chain of activities of zero total float should appear very clearly on the schedule.

A lack of Critical Path shows that the schedule is not properly linked (in particular regarding the end activities of the Project) and probably uses too many internal date constraints. In any case it is not a sound schedule to be used for decision-making at Project management level. Request the planning team to revise.

What Needs to be Checked on the Schedule Source File

When all the previous checks show positive results, further investigations can be performed on the schedule source file. This requires looking at the schedule in the original scheduling software (Oracle Primavera, Microsoft Project etc.) so as to observe the actual dynamics and linkages. It does not require specialist training, just some practice and logic and it is absolutely fine to get the software manipulated by the planner during a workshop, or a conversation (in all cases the planner will have to implement the changes so it is better that he/ she is present).

Some specific software can perform automatically a health check of the schedule and can be used as a support to the following checks. PERTMaster (now known as Primavera Risk Analysis and integrated within Oracle Primavera division) includes such a function (called "schedule check"), as well as Acumen (recently bought by Deltek). Both softwares were developed by the same group of people. However it is absolutely not a requirement to run these automatic software checks to assess the quality of the schedule. It is a plus if they are available, because they give an exhaustive list of the activities in the schedule that have strange properties (missing links, start-to-finish links, etc.).

The main point is to check that the schedule effectively reflects a fully-linked, free-moving network of activities with the appropriate linkages and the minimum of constraints.

The main points to check

Some useful heuristics and key issues:

- In most sound schedules, the number of links should be around twice the number of activities (on the order of between 1.5 to 2.2 times). Less links indicate a weakly connected network with possibly insufficient representation of internal constraints between activities and chains of activities; more reflect an overcomplicated schedule that might have been excessively linked and thus constrained,
- Open-ended tasks should be of course avoided as all activities should have at least a predecessor and a successor, except milestones, the start of activity chains and the final activities of the Project,
- The number of date constraints (where the start or the end of an activity is constrained by a date) should be minimal. They should be replaced by schedule logic. This also applies to soft constraints (such as 'start later than…'),
- Start-to-Finish links should be avoided (and they won't work for probabilistic risk analysis), except possibly for milestones links.

More advanced issues include:

- Avoid negative lags. A negative lag is an overlap in the logic between two activities – often it is used to represent an activity starting earlier, with sufficient time allow some other work to happen. Lags cannot have risk or uncertainty in Schedule Statistical Analysis. In reality it is likely that the negative lag represents a necessary overlap, whose duration is uncertain. Consider replacing a negative lag with another kind of link that does not need the lag. For example, replace a negative lag on a Finish-to-Start link with a positive lag on a Start-to-Start link; or split the activities so that the overlap is explicitly represented by an activity.

How to run a schedule source file check

In summary, play with the schedule to check that it flows properly and without hurdle!

The checks should be done with the schedule running live on the screen, and the same representation used as for the checks on the schedule print-out (in particular, with the Total Finish Float apparent).

- When **changing significantly and artificially the duration of an activity** (as a test, double or triple the duration of a key activity to see what happens), the full network should flow and change without visible artificial constraints. It is quite easy to see visually on the Simplified Project Schedule and can be a bit more difficult to see in the Integrated Project Schedule, but it can still be done,
- When significantly and **artificially lengthening the duration of an activity** that is not on the initial Critical Path, alternate Critical Paths should appear running through the entire schedule duration,
- **Follow the logical links from one activity to the next, in particular on the Critical Path**. It is easily done in all scheduling software through the function that identifies all predecessors and successors of a specific activity, and allows jumping directly to these other activities. Through a suitable sampling, this will quickly give a sense-check of whether the EPCC (Engineering- Procurement- Construction-Commissioning) chains are properly represented, as well as whether the most important interfaces and dependencies between chains of activities are properly linked (a Project Manager will generally know what these are). In that review it is important to keep it at the sampling level for the Integrated Project Schedule without following on to review the entire schedule. If the sampling on 25 key activities and expected linkages, and on 5 main E-P-C chains fails often, there is a strong possibility that the schedule will need to be reviewed thoroughly. If only a few misalignments are noted, the schedule can be expected to be fine overall. On the other hand, this review can be expected to be done in an exhaustive manner in the Simplified Project Schedule: this is also required if the Simplified Project Schedule is used for Schedule Statistical Analysis.

- Sample activities and **look for the number of successors and predecessors**. Too many predecessors and successors, in particular if they use the same resources, is unrealistic. Also, this tends to effectively reduce the float available in the schedule network and is hence diminishing schedule resilience (see the concept of 'merge bias' in Chapter 7).

Proper schedule coding to produce different views

The final check point is to check that the schedule is properly coded so as to enable easy filtering and display in ways that will be useful to different Project contributors.

Custom fields should allow filtering/ display the schedule by the Work Breakdown Structure dimensions (refer to appendix 5):

- Function/ Trade (Engineering, Procurement, Fabrication, Construction, Commissioning, etc.),
- Plant/ facility area,
- System,
- Contractor (when relevant).

Each of these needs to be further subdivided in sub-functions and sub-areas as useful.

Should these codes be missing, the schedule will not be very useful for the team. To draw maximum utility, this coding should be introduced, and a number of pre-defined schedule views setup so as to allow automatic production of all the relevant and appropriate printouts on a regular basis.

Proper schedule coding for Earned Value Management

In addition, in those areas where Earned Value Management is planned to be implemented (typically, engineering and construction), specific coding needs to be implemented. This coding has to be entirely consistent with the Work Packages used by Cost Control. This breakdown structure needs to be implemented in the schedule in the form of a separate specific field so as to be able to implement Earned Value Management according to the relevant subset of the WBS.

Schedule Realism Checks

Once Project schedules have been checked technically and logically, it is important to also check the soundness of the information that is included, with regard in particular to the estimated duration of activities.

This requires the knowledge from people of the trade. In mature Project organizations, databases of actual data from past Projects will be available for benchmarking, possibly with some formulas for scaling when relevant. Suppliers and contractors will include time estimates in their bids that can be used as a basis for parts of the schedule (however, they need to be carefully challenged because delivery lead time is a competitive parameter, hence those proposed durations need to be reviewed against benchmarks and the inclusion of an additional allowance is often necessary). Finally, for activities that have never been carried out before, expert knowledge can be mobilized and documented (as well as the related uncertainty).

In this section we will document first the main traps of estimating; and second, some useful ways of looking at schedules to check their realism.

Main Traps of Estimating

Failure to document the estimates' source

It is essential in all cases to be able to relate each time estimate in the schedule with a particular source of information and document it. It is an area where proper documentation of the Project preparation stage (or tender stage for a Contractor) is necessary. First, large Projects have long durations and the people who will run the Project a few years down the road might not be the same as those who prepared the Project; and in any case such documentation is important on contractual and traceability grounds.

Failure to challenge the experts' durations

We will call 'experts durations' all those durations that are estimated by people without the support of hard data evidence from past Projects. This will for example apply to all estimates of engineering activities except in the most mature organizations where detailed benchmarks will have been developed over time.

Because people tend to feel judged on the durations they announce as a sort of commitment, they will tend to pad their estimates, i.e. announce duration estimates that will be longer than what they could expect to be achievable.

In other cases of experts' estimates for activities that have never been done before, estimates are generally found not be very well calibrated and to be too optimistic. In general, the less detailed or known in detail the activity to assess is, the more optimistic estimates will be.

It is thus critical for the Project Manager to be sceptical with all estimates arising from other sources than hard benchmarking data, and to take the time to challenge the information. This is more or less feasible depending on the relative weight of functions and Projects in the organization; experienced Project Managers will know where to add some Project-specific contingency and where durations can be lowered to challenge the teams. External peer reviews are also very useful practices in that particular case.

Issues with schedule resourcing

Project schedules are often 'resourced', at least for some sections that involve internal resources or resources directly managed by the organization. It can also be done for construction activities when good quality benchmarks are available (for example in man-hours per ton of steel or reinforced concrete).

Resourcing of schedules is a best-practice in particular to check the feasibility of challenging schedules in terms of duration. It allows the Project team to defend overall schedule durations when over-eager executive management would tend to try to raise expectations of an early finish.

Schedule resourcing raises the following challenges:

- **Resourcing needs to be done based on proven benchmarks**, with an adequate breakdown by type of resources. This breakdown needs to be done at a level that allows for analysis but not too detailed to become difficult to maintain and overview. As a best practice, a maximum of 10-15 trades should be implemented for the relevant macro-phases of the Project, including the few critical trades even if the number of resources involved is relatively limited. Depending on the industry and the Project setting, these critical trades are easy to identify.

- **Resource-levelling in case of resource constraints needs to be treated with a lot of caution and needs to be done manually.** It is dangerous to rely on the automatic routines built in the scheduling software. First, they won't give the same result depending on the software and the software will be blind to actual Project priorities. Hence, resource levelling needs to be done manually by moving the impacted activities in the schedule to limit the resource usage peaks, and look at the resulting resource utilization curves.
- **If you don't change your project completion date, resource levelling due to resource availability limitations will diminish the overall available float.** You need to check that the remaining float is sufficient and that you have not created too many marginally critical chains that will make the expected project completion improbable. In general, resource levelling needs to be accompanied by a relaxation of the project completion date.
- **Adding resources does not necessarily imply a significant shortening of Project duration.** Adding resources on Critical Path activities will make them shorter, but will quickly highlight the influence of sub-critical sequences of events, and will shift bottlenecks to the activities that follow immediately. There is also rarely a proportional relationship between the number of resources and actual productivity due the loss of productivity in large teams or when several shifts are implemented.

Useful Check Points for Project Schedules' Realism

Checks for the entire schedule duration

A Project schedule should fit in the existing benchmark of similar Projects. This seems obvious but is often overlooked because of micro-analysis of the schedule, and the fact they are often unfortunately built in a bottom-up manner. In a given industry where similar size facilities generally take 48 months to build, it is quite unlikely that a new facility that includes some cutting-edge technology that would be implemented for the first time would only take 40 months to build – unless there is an actual straightforward explanation to that.

Project Managers generally have a sense of what can be achieved in a given industry. Still we have observed situations where for commercial reasons Projects were sold with durations lower than the reasonable benchmark for the industry, putting the Project Manager from the start in a very uncomfortable situation (in these cases the Owners are also at fault not to have compared offers to existing benchmarks for similar facilities).

Another way to check the entire schedule duration is to examine the Critical Path and check what actually drives it. The issue will be different whether it is the procurement of long lead items, the engineering phase or the construction phase that drives the schedule. A sense of the required duration for the Project can be obtained from this analysis. This also requires getting a feel of how sub-critical the other activity chains are and how possible it would be that they would drive the schedule (refer to Chapter 7 on Schedule Statistical Analysis).

Checks for key holiday periods

Well-built schedule calendars sometimes include public holidays at the Project team's location, but generally do not cater for those periods where activity is notoriously low in a given country or region, while a significant part of the procurement or construction scope would be done there. International Projects nowadays have global supply chains and they often stumble on these key holiday periods, as sometimes their schedules expect critical activities involving a lot of local resources to happen independently of these low-activity periods.

Common examples of such key holiday periods include:

- Summer period (mid-June to mid-August) in US and Europe,
- Christmas/ New Year period in Christian countries;

- Chinese New Year in China and some parts of South-East Asia (depends on the lunar Calendar, between end January and end February) (for construction work, this period is compounded by the difficulty that a lot of cheap construction labour going home for the holiday won't return, creating a slow restart and the need to re-train workers)
- The Ramadan month (slow activity) and the End of the Ramadan festivities in the Muslim world (variable period in the year)
- The monsoon season (July-August in India, November-December in South-East Asia) for all outside works,
- The carnival period in Brazil,
- Punctual events like the soccer world cup in Latin countries,
- Etc.

This issue requires some research based on the location of the main contractors and worksites so that these holidays can be fully taken into account and the temporary lower activity do not come as a surprise. We recommend that these key periods be identified in the Project schedule. They will sometimes act as a constraint to make sure parts of the work scope are finished before those festive seasons. We are always amazed how finely developed schedules forget about these issues and the amount of disruption it can create in Project execution.

Checks for schedules that are resourced

When schedules or parts of the schedules are resourced, additional checks can be run:

- Of course, **resources utilization vs overall resource availability**. This needs to be checked overall, and also for critical resources. Critical resources can be for example:
    - People, generally specialized resources such as:
        - Qualified welders,
        - Non-destructive analysis personnel,
        - Crane drivers...

  - Specific supplies, such as:
    - Concrete for concrete-intensive construction work,
    - Bulk material that poses logistics challenges,
    - All specific critical equipment, such as earth-moving equipment for earthworks.
- **Resource density limitations** need to be worked out to understand the maximum resources that can be deployed on a plant area; and then check that the Project schedule does not exceed these density limitations for each area. This needs to be done on Projects that are constrained in space such as shipbuilding, FPSO construction, process plants construction etc. There can be additional constraints related to the different working heights, which would not allow for simultaneous work on different elevations. The best practice is to split the work area in about a dozen areas, work their area / volume and the maximum reasonable density of resources in each area; whether there are some particular simultaneous work constraints; and then check from a schedule download how the Project schedule fares. If that cannot be done, simple considerations about the peak number of workers will often give a good idea of the reasonableness of the schedule.
- **Resource mobilization and demobilization curves** are also a great indicator of the reasonableness of the schedule. Project schedules that have been worked out carefully will often pass the previous tests but fail this particular test. It is unrealistic to mobilize or demobilize personnel too fast on a Project. There are logistics constraints; requirements for permits and trainings before starting on the job; availability of the required tooling; learning curves for all newcomers, adjustments to site supervision etc. Each industry has some benchmarks as to reasonable mobilization rates that need to be further challenged taking into account the logistics for remote areas. Mobilizing overall more than 15% of the peak construction personnel on a remote worksite every month is not realistic. Historical effective mobilization data will be available for fabrication yards, shipyards etc. In addition, there

will also be limits for specific trades and mobilization/ demobilization curves also need to be considered carefully. It is not realistic to expect excessively quick mobilization, alas many Project schedules and most 'acceleration schedules' do fail this elementary test which is a physical limitation of all human endeavours.

A word about the tendency to shorten fabrication or construction works by working double shift. Except where this type of work schedule is explicitly already organized and proven as a 'normal' organization mode (e.g. offshore construction vessels, some fabrication facilities), this should not be considered as a baseline for the schedule. The use of double-shift as an acceleration measure should be considered only carefully: having two teams that succeed to each other on the same job invariably generates handover issues, productivity losses and quality non-conformances. In addition:

- on all sites that have not been purpose-built for double shift work, maintenance and housekeeping still needs to be done, which requires some time,
- some key bottleneck equipment might already be used 24h per day (cutting, automated fabrication, etc.),
- Some activities cannot be accelerated physically (e.g. concrete curing) or are required to happen by day-light (e.g. harbour manoeuvres, specific lifts),
- on most construction sites, there needs to be some time with very limited personnel on the premises for gamma-ray non-destructive testing.

Conclusion

We are always astonished to be called in to review Projects only to find out that they have, for starters, a poor schedule. Poor in the sense of poorly linked, not representative of the work to be done, or unhealthily unbalanced between the types of activities that have to be performed.

Setting up an adequate route map at the onset of the Project should be the utmost priority of the Project Manager. Unfortunately, this does not always happen, either because of work overload or of a lack of competency from the Project Manager. This Chapter provides Project Managers and other senior Project personnel with straightforward ways to challenge a schedule to get it improved to a point where it can be realistic and useful.

In all cases, before sailing away, make sure to have a proper map of the right quality in hand!

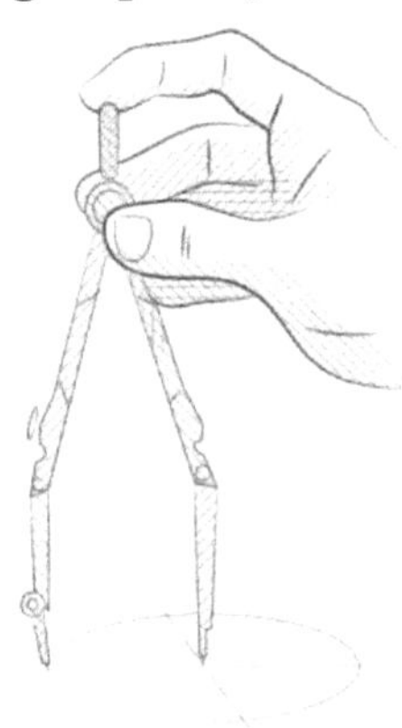

Chapter 7: How to Increase the Resilience of your Schedule: Schedule Statistical Analysis

Note: we reproduce here for the sake of completeness the same Chapter from our Project Risk Handbook with some slight adaptations.

Chapter Key Points:

- To be performed properly, Schedule Statistical Analysis needs to be applied to a simplified, fully linked schedule that correctly represents the drivers of the Project (the Simplified Schedule of our schedule hierarchy).
- The probabilities regarding the Project delivery date are what is often requested, but it is not what is the most interesting when performing a Schedule Statistical Analysis.
- Schedule Statistical Analysis allows examining which activities are critical with a high probability.
- Merge bias, where several chains of activities compete for criticality and converge at some point, is a major issue regarding schedule resilience and must be corrected.
- Improvements to a schedule after a Schedule Statistical Analysis should rather focus on improving the schedule resilience than trying to shorten it.

Introduction: the Straightest Way is not Always the Fastest and Surest

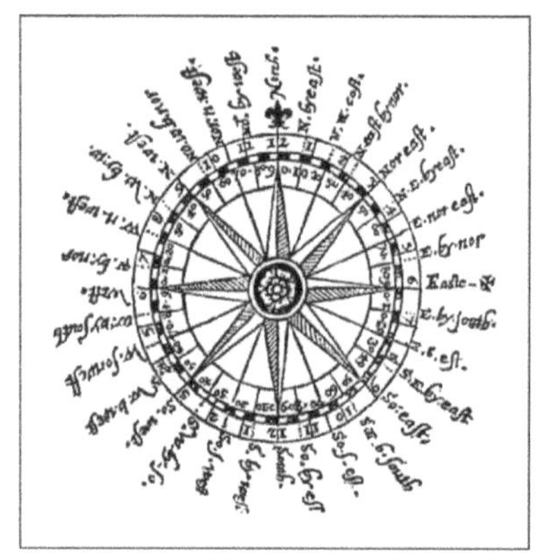

When planning a voyage, the first tendency is to seek the shortest way on the map – the straight line for short distances, the major circle on the globe.

Old sailors knew well that it wasn't generally the surest and fastest way. Taking into account the uncertainties of weather, it was best to plan longer journeys distance-wise that could benefit from the continuous supply of trade-winds around the tropical latitudes. By following this practice, the duration of the journey would be more reliable and the risks to be drawn to unknown shores by an unexpected storm much reduced.

As we will examine in this Chapter, schedule optimization should not necessarily seek shortening the distance by shortening the Critical Path. It is often much better to work on improving the resilience of the Project schedule to diminish the uncertainties of Project duration and build resistance to unexpected external events.

What You Should Really Seek When Conducting a Schedule Statistical Analysis

Schedule Statistical Analysis (SSA), also often called Schedule Risk Analysis, aims at understanding the impact of variations in the activities durations on the overall Project outcome. It is a bottom-up method where the possible variations on all relevant activities are input into a model which then uses a Monte Carlo simulation to predict statistically the Project schedule outcome. In the model, the activities are linked logically so that delays on an activity will delay the start of the following activity, etc.

Most of the time, the prime motivator for management to request for a Schedule Statistical Analysis is to have an idea of the statistical distribution of the finish date of the Project. This is often important for contractual and investment justification.

However we argue that this should not be the main objective of such an analysis. A SSA can bring much more value in a number of areas than just an estimate of the finish date, which will be inaccurate anyway (more on that in the section on limitations). Actual value to be drawn from the method include:

- Generation and utilization of the Project Simplified Schedule showing the actual drivers of the Project, that is a very valuable asset for communication to stakeholders and induction of Project contributors,
- Identification of the Critical Path/ chain and measurement of the actual sub-criticality of other sequences of activities to give a good idea how to drive the Project execution and avoid discontinuities in criticality,
- Identification of the sensitivity of critical interim milestones and convergence points,
- Improvement to the schedule resilience by observing criticality ratios and by testing 'what if' scenarios and observing their impact on the Project execution.

How to Perform Properly Schedule Statistical Analysis

It does not make sense to perform the SSA on the basis of the Integrated Project Schedule covering thousands of activities, for the two following reasons:

- It is impractical to have meaningful statistics on the variability of thousands of activities; in addition people generally can respond to questions relative to aggregates, not detailed activities;
- It is not appropriate from a mathematical perspective (as multiplying the number of activities in a Monte Carlo simulation automatically

diminishes the overall variance – refer to our Project Risk Handbook).

The first step – and the most important, requiring the most work, is thus to devise the Simplified Project Schedule, linked logically with aggregate activities. This cannot just be a roll-up of the overall schedule; it needs to be carefully crafted to reflect the actual Critical Path, the subcritical chains of activities, and in general, reflect appropriately the high level logic of the Project execution. To achieve this, people familiar with the Project and the schedule need to be involved. Refer to Chapter 2.

Only when this is done, can data be sought regarding the possible variability of the activities of the Simplified Schedule. It is rare that companies have hard data on the matter and often, it is necessary to resort to professionals' experience. A word of caution – professionals will tend to underestimate the variability brackets unless appropriate questions are asked like 'have you ever seen this happening'. This is due in part to their necessarily limited experience, which needs to be complemented by company-wide and industry-wide lessons learnt. A proper calibration of risk estimates has thus to be done prior to the exercise. In particular, the possibility of substantial deviations from the deterministic durations might not get mentioned when they actually do occur in reality (e.g. doubling the initial duration of an activity).

An Opportunity & Risk register can be applied to the model to include a number of discrete risks and scenarios to the simulation; however this additional complication is not always needed.

The final part of the SSA, running the simulation using a tool such as Primavera Risk Analysis (or some similar tool) is actually the simplest. Mathematical consistency needs to be checked to ensure a sufficient number of iterations have been used for the Monte Carlo procedure.

The Limits of Schedule Statistical Analysis

Like any model of reality, SSA has a lot of limitations and its results should absolutely not be taken for granted. It has the advantage over straightforward Monte Carlo procedures like the ones used for cost to take into account logical relationships between activities. However, there are still a number of issues that are not taken into account, in particular:

- Common/ shared resources used by several activities, actual resources limitations and the related snowball effects of poor performance on a particular deliverable on resource utilization and Project management focus,
- Low probability, high consequence risks are poorly modelled (and often ignored in the models) whereas they can have significant importance in real life Projects,
- The model does not consider any local change of logic and change of sequence of activities, whereas this is what happens in the reality of Project execution when Projects try to mitigate delays on certain activities of the Project.

It is thus important to underline that in spite of its very attractive approach and presentation, SSA can absolutely not be taken as a full representation of reality. It is only a model that shows the effect of certain events but is necessarily much simplified.

In particular, Projects in reality very often show delays which are much greater than whatever had been found using SSA due to real-life cascading consequences of delays in terms of resources utilization and exhaustion; or the occurrence of low probability, very high consequence events which cannot be modelled by the method. In fact, for SSA results to be closer to reality, an active Project Management is implied where small variances and delays are actively managed and compensated on an ongoing basis (in the spirit of the Convergence Plan where key gates are defined with fixed dates).

What Results Should be Analysed?

The overall result of the expected delay to the Project should thus be considered with much caution and should not be supposed to represent reality. It can be used in a number of processes, e.g. as a contingency or buffer that can be added to the schedule (ref. Chapter 11). Unfortunately, a lot of organizations only focus on that particular outcome and not on what is really interesting in a SSA data.

SSAs provide a raft of results for each activity in the schedule being modelled, including distributions for its start, end, float, how the duration of that particular activity is correlated to the overall Project's duration, and what is the probability for that activity to be critical.

All this information can be useful, still we believe this last parameter is one of the most interesting overall. What is really important is to understand how robust the Critical Path of the deterministic schedule really is. It happens often that this analysis uncovers that actually the Critical Path is only marginally critical and that there is a number of other paths that can become critical with a high probability (sometimes over 50% of the Monte Carlo trials; any criticality probability over 20% is a concern). This is an important issue that needs to be tackled.

Conclusion on Schedule Statistical Analysis

'Schedule Statistical Analysis' (SSA) is done more often because it is a mandatory exercise than because people find it useful. And when it is done, in most cases people focus on a result that is quite meaningless – the Project finish date statistical distribution. This is unfortunate because when properly conducted, Schedule Statistical Analysis can bring a lot to the understanding of the Project schedule – and what are its actual drivers. In particular, Schedule Statistical Analysis can be used to identify improvements that can significantly enhance the robustness of a Project schedule, as we will now explore.

Tips for Enhancing Schedule Resilience and Project Manoeuvrability

One of the most important things to know when you execute a Project is to know where your current constraint (or bottleneck, critical chain) to Project delivery is. It is important that it remains stable and clearly identified during Project execution. Hence, it is preferable that the critical activities remain critical, and that the Critical Path does not jump unexpectedly from one side of the Project to the other.

Often, the first focus of management will be on trying to shorten the Critical Path to get the facility to operate earlier. This is not necessarily the right thing to do because it increases the risks of delays compared to expectations. By doing so, one inevitably raises the probability of other chains becoming unexpectedly critical. The question is whether it is better to have a schedule with a very neat Critical Path and other chains of events very subcritical, or it is better to have a schedule with many chains of events competing for criticality?

Having at all times a very clear Critical Path for the Project (and all other chains of activities very sub-critical) is a great asset that makes the schedule much more resilient, for the two following reasons:

- In terms of decision-making and management focus, it is far better to be sure at all times where the Critical Path is and will remain. It gives a direct handle to Project management on the actual delivery of the Project, and a sounder decision-making platform. Most Project surprises (and outright failures) stem from the fact that the actual critical activities have shifted and it has not been recognized by the Project team in time;
- In terms of resources management, deep sub-criticality of all the other chains of activities in the Project will allow to redeploy resources to the critical chain so as to compensate unexpected issues and events, without changing the overall logic of the schedule. This creates a much higher level of

intrinsic resilience, as surprises can be compensated by re-deploying resources internally to the Project.

Hence, before trying to optimize the schedule duration by optimizing the Critical Path it is much better to check the overall resilience of the schedule and ensure that it is up to the minimum that would be required from the Project perspective.

This is why observing the resilience of the Critical Path in the SSA is possibly the most important result.

Case of Parallel Similar Workstreams Competing for Criticality

If the observation of criticality of different chains of activities shows that several work streams are competing for criticality to the point of creating a situation where the Critical Path can jump easily and unexpectedly, action needs to be taken to improve the Project plan. It will enhance significantly the schedule robustness and the ease of execution.

Having several chains of activities competing for criticality significantly increases the probability for the Project to be late, because any delay to any of these chains will impact the Project completion date; and the other way it will be very difficult to accelerate the Project because such acceleration would need to happen in several parallel activities at the same time. In addition these effects are reinforced by the fact that those parallel chain of activities often share the same resources.

This issue is enhanced when those critical chains have to meet at a convergence point. In the SSA literature this effect is called "*merge bias*". The risk at the merge point (convergence point) of two or more activity chains is greater than the individual risk of any of those chains. This effect is significantly reinforced when those chains are all close to being critical. The Convergence Plan is an essential tool to address these issues.

Hence in any situation where two or three similar chains of activities happen in parallel and compete for criticality, removing this competition (e.g. by starting some chains earlier, and not the critical one) provides great benefits:

- Less strain on resources if similar resources are used,
- Resources used on very subcritical chains of activities can de facto be used as a buffer to add resources to critical activities that would need such support, without creating a situation where the Critical Path would jump somewhere else,
- The learning curve will happen on non-critical activities instead of impacting directly the Critical Path,
- If these chains converge together, the convergence point will be better protected from events and fluctuations.

This is shown on the figure below:

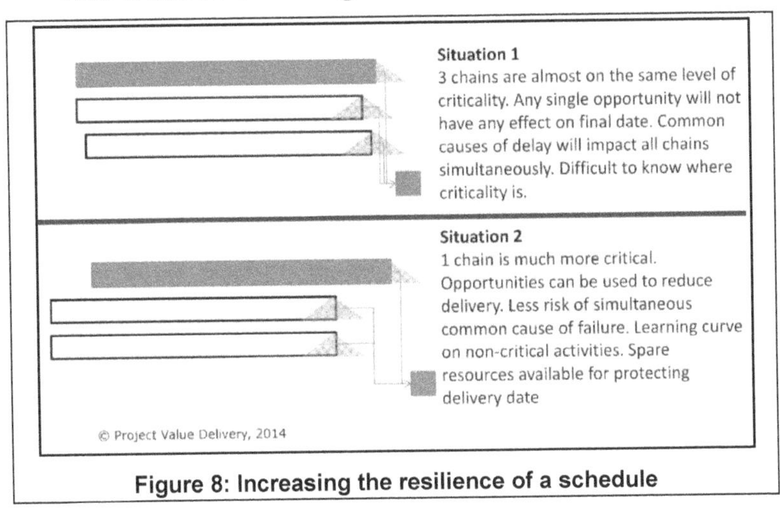

Figure 8: Increasing the resilience of a schedule

Additional Hints for Increasing Schedule Resilience

Depending on the circumstances of the Project, there are other ways to increase resilience:

- If the Project consists of several chains of events that are realized by the same resources (e.g.: several modules on a yard, or several areas of a plant), there is also a great benefit to have the non-critical chains of activities start earlier than the critical chain, if possible. This will ensure that the uncertain learning curve that is inevitable at the beginning of any type of activity will not impact the Project's Critical Path: the subcritical chains of activities will have borne the delays and the risk, which is fine as they are very subcritical. Even substantial difficulties will not impact the Project delivery date.
- If the Project contains a chain of activities which is very uncertain, and thus could unexpectedly become critical and drive Project delivery (e.g. a component particularly difficult technically or requiring a substantial R&D effort with high uncertainty), it might be useful to plan to have two or more solutions competing up to a certain point. In some instances it might even be required to order twice similar critical parts from two different suppliers (thus ending up with a set of unused parts) just for the sake of containing the risk to the rest of the Project within acceptable bounds. The additional expense can be seen as an insurance.

The Investment in Schedule Resilience

This last example illustrates clearly that increasing your schedule resilience has an implied cost. In the choice between trying to shorten as much as possible the Project delivery and rather, increasing resilience by increasing the sub-criticality of non-critical chains of activities, cost and time will come into account.

It is one of these typical situations where organizations dealing with Large and Complex Projects will know how to invest in decreasing significantly their risk, contrary to most organizations that will only try to minimize cost (and maximize promised but elusive return on investment).

The issue is not to come up with the shortest schedule possible and kick off the Project with an unrealistic expectation, but rather to ensure that Project delivery will happen as expected notwithstanding the inevitable disturbances from the Project's environment. A resilient schedule is far preferable to a shortest schedule where most activity chains will compete for criticality.

Luckily most of the time when resilience is being analysed on a Project schedule, there are ways to improve it without impacting the Project's overall schedule, which is the main cost driver. It leads to the counter-intuitive changes of starting earlier the sub-critical chains of activities to improve significantly the Project's overall resilience. That type of change is often easy to implement with a minimum cost impact.

Conclusion: Improving your Schedule Resilience before Starting your Project is a MUST

It is now part of Project Value Delivery's methodology to examine the resilience of a Project schedule by running an appropriate Project Schedule Statistical Analysis and examining the criticality probability of the different activity chains.

It is often possible to improve significantly the schedule's resilience at minimum cost by some counter-intuitive actions such as starting sub-critical chains earlier.

The advantages of this practice are numerous; the most notable arguably is to give the Project management team a stable focus on a Critical Path that will not change except in case of major disruption, hence ensuring appropriate decision-making throughout the Project.

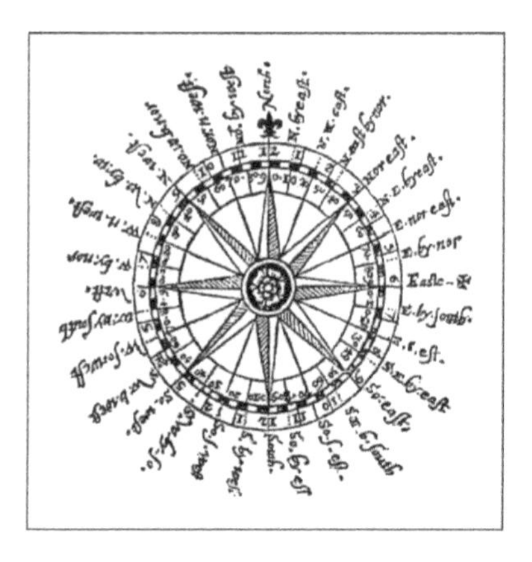

Chapter 8: Updating the Schedule for Actual Progress

Chapter Key Points:

- Proper schedule update is required to know the position of the Project and this update needs to be accurate.
- It is easy to determine whether an update is correct by asking the people doing the work on the ground and comparing their answers with what is reported in the schedule.
- One issue is sometimes the fear to face reality, but a schedule update must reflect actual reality.
- The fundamental process is proper communication. Planners need to go out and get the required information. Conversely, Project team members must be conversant with the schedule and inform the planners in case of changes.
- All changes to the schedule including re-sequencing and changing sequences must be traced formally.

What Schedule Updating is About

The previous Chapters were about producing a useful map for our expedition – with the right scale, level of details and consistency. In this Chapter, we will deal with Updating. This process responds to the first fundamental navigation questions:

- Where are we currently?

In the old days of sailing ships, and before a reliable chronometer was invented in the late 18th Century, figuring out one's position at sea was a delicate task. Most of it was an estimate by dead reckoning - based on the observed speed and heading of the vessel. Because it did not account for leeway or current drift and the adding up of measures of speed and direction were somewhat unreliable, in particular when the ship was tacking to progress in the wind, navigation remained quite unreliable once one had left the comfort of seeing the coast. Arriving on the other end of the ocean, it was always a challenge to identify the whereabouts of the actual landing point and avoid the associated dangers, because the uncertainty on one's position did increase with time, until a physical observation could once again diminish the uncertainty.

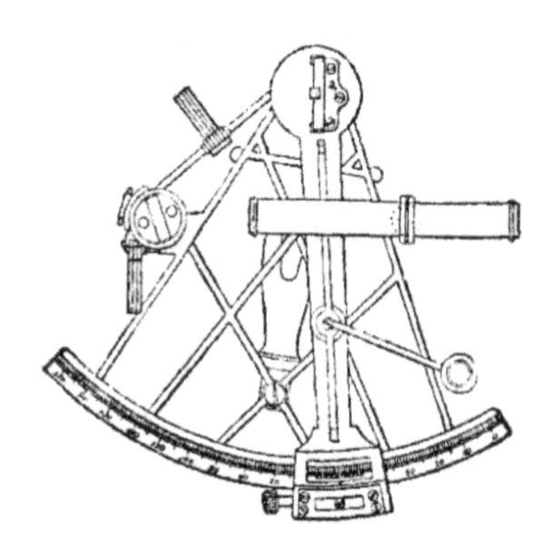

In the 19th century, with the chronometer and the sextant, navigation was still estimated by dead reckoning for 24 hours and every noon time, a precise observation of the sun would give a position, which would then be the start of the next 24 hours period (this process was called '*taking a sight*'). The quality of navigation was highly dependent on the quality of the daily *sight*. The inaccuracy of the measurement directly translated in an inaccuracy in position.

For practical reasons, all Projects nowadays do navigate like in the 19th century: a full update of their schedule chart is only done every week or month, and in-between, navigation is done by dead-reckoning. The issue is, how precise is the periodic position measurement? And from there, how good will the forecast be for future navigation?

Let us not forget that although basic and simple, this process is not implemented satisfactorily in many organizations and Projects, leaving the possibility to hit shoals that one would not expect to meet on his route while they are properly mapped and well known. It is therefore important not to take it as a given and spend enough attention to make sure that it is being applied properly and consistently. In the good ol' days even if junior officers were taking the sights, the captain made sure he would check the position calculations!

In the following sections, we will concentrate on the update of the Integrated Project Schedule. The lower level detailed functional schedules are often updated through systems with the help of personnel to check the accuracy of the data. The Convergence Plan and the Simplified Schedule can be updated based on the Integrated Project Schedule's data.

Schedule Updating Needs to be a Bottom-Up Exercise

The Integrated Project Schedule (and the more detailed functional schedules) needs to be updated in a bottom-up approach, activity by activity.

It would not be effective to try to update by group of activities or at a high level because it is not the level at which the update information is available.

Actually, many of the Integrated Project Schedules' activities may not be updatable directly, but only through the aggregation of more detailed updates available in the detailed functional schedules.

How to Quickly Check the Quality of a Schedule Update

During our consulting assignments we are sometimes astonished to observe that the periodic update of the Integrated Project Schedule is not accurate at all. This is easy to ascertain: the situation on the ground is quite different from what is described in the schedule!

This generally occurs in organizations where planners are considered more to be 'scheduling software operators' than people that can bring value to the Project by fostering communication. This is a serious mistake, encouraged by the fact that many 'planning engineers' on the market effectively sell themselves more as 'scheduling software operators' without necessarily understanding the particular business branch.

What needs to be understood is that planners can, and do generate a large part of communication flow in the Project, by getting information from all contributors, merging it in a meaningful framework that then allows the Project Manager to identify and tackle those issues that prevent smooth execution. Planners also feedback information to contributors through their schedule updates (ref. our *Project Control Manager Handbook* and the concepts of communication and data assurance).

It is very simple to check the quality of a schedule update: take the latest update of the Integrated Project Schedule and confront it, both in terms of present situation and short-term forecast, with people doing the groundwork. It is enough to do that on a sample that covers all types of activities (engineering, procurement, and construction). You will very quickly ascertain:

- Whether people in charge of the work have been interviewed and feel they have been listened to when it comes to the work progress and short-term forecast,
- Whether the schedule actually reflects the real situation as felt by the people that are close to the Project execution.

It is not generally even needed to go and check the physical progress on site, unless there is a significant doubt of inadequate reporting for other reasons. Most often, people in charge of construction don't disguise reality.

Such checks are easiest done by someone from outside the Project team to ensure candidness and avoid conflicts of interest; but it can also be done quickly from inside the Project to get a feel of the quality of the information.

The Problem of Facing Reality

Sometimes we encounter Projects or organizations that maintain a delusive schedule, either by not updating it, or by only updating certain activities that are not really critical. It can be encouraged by certain organizational cultures where candidness is not appreciated and cover-up recommended (these organizations often show a very high rate of personnel turnover).

This is an absolute killer when it comes to Project execution. Reality might not be nice to look at, but it certainly needs to be stared at in all its dimensions so that the right decisions can be taken at the right time. Reality is not scary in itself; fear is only an emotional reaction to reality.

One should not be scared to be candid when updating the schedule. Reality is what it is. The role of the Project management is to deal with it.

Schedule Update: Properly '*Taking the Sight*' of the Project

What are the basic health recommendations that need to be followed for getting a truthful picture of the Project condition?

Distinguish between Precision and Accuracy: Seek Accuracy First!

Before we delve into the failure modes and root causes of unreliable snapshots of the current Project situation, let's remind ourselves of the important distinction between accuracy and precision from the theory of measurement (already mentioned in our *Cost Control handbook* on cost forecasting):

- Accuracy is how close a measurement is to the actual value,
- Precision is how consistent the measurement is if repeated many times (but that does not mean that it is accurate with regard to the actual value).

A measurement system can be accurate and not precise, precise but not accurate, both or none. It is very onerous to have a system that is both precise and accurate.

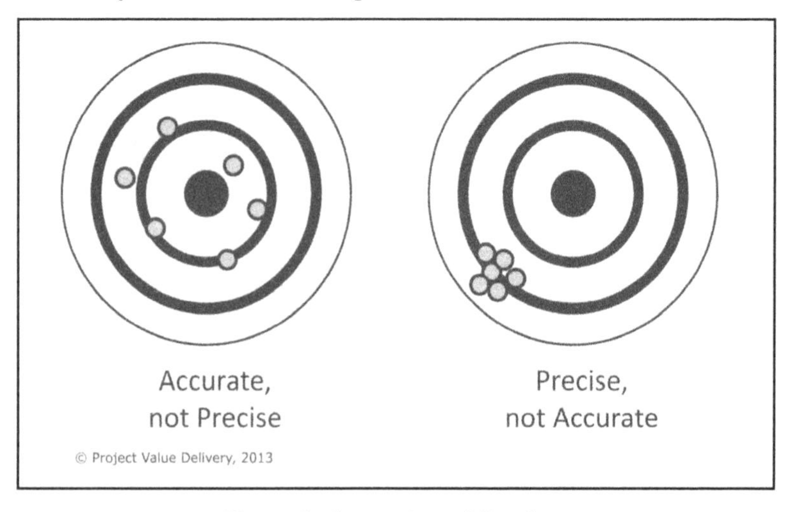

Figure 9: Accurate and Precise

As we measure the current situation of the Project, we mainly aim at accuracy. While precision would be a bonus, we won't repeat the measurement too many times for each update; at the next update the error can be expected to be randomly moving, and average out eventually. In addition, total precision is not necessary in the dynamic context of a Project. So, when it comes to Project snapshots, we seek accuracy first. Precision is not so important – that's why the common practice to extrapolate progress for the last few days of the reporting period is not too much a problem. Any imprecision will be compensated in next month's update.

It would be the same if we were 'taking sights' in the middle of the ocean. What counts is the accuracy, so that dead-reckoning errors do not add up. We are not too concerned by precision because tomorrow's measurement will compensate and average out.

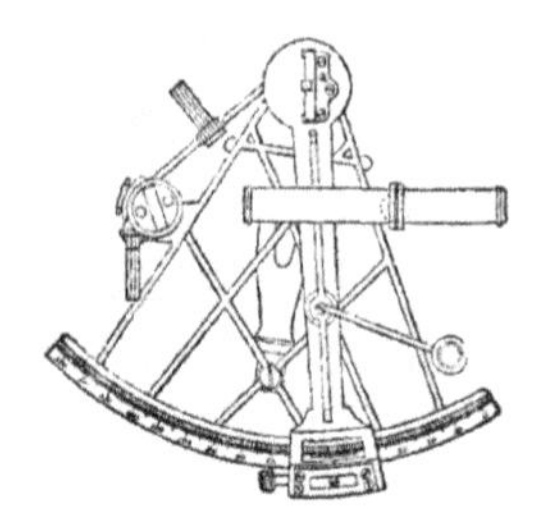

Updating the Project Situation is About Mobilizing the Common Knowledge of the Team

Updating a Project schedule is not a solitary exercise; it requires mobilizing the knowledge of the Project, which is spread among its members. Because Large, Complex Projects involve large teams that can be spread over several locations it is not necessarily an easy thing to do.

The difficulty is increased by the need for planners to be sometimes assertive to get information about the latest status and perception of progress from the Project team members (this difficulty is compounded by the fact that many experienced planners are introverts).

Still the underlying process is one of the rare processes that aims at collecting information spread around in the organization, structure it in a consistent way and roll it up to give a clear information about the Project status.

Some organizations will rely on comprehensive automated systems as a foundation for schedule update, typically systems covering the activities of the main departments. However, having a comprehensive system is not a guarantee for correct data. Our observation as consultants is that Project-driven organizations that rely excessively on systems (and generally, at the same time, tend to over complicate their schedules and all tracking databases) actually suffer from a much lower update accuracy. This is because they over-rely on systems without realizing that the quality of the information is what is being keyed in by Project team members. And by making team members update numerous systems without them understanding the importance of that work, those projects expose themselves to large inaccuracies and out-of-date data as this updating task will often not be a priority.

Schedule Update Failure Analysis

Failure modes

What are then the failure modes of this information collection process? We seek complete, timely, accurate information covering all Project activities. Failure modes can thus be classified in the following four main categories (which can thereafter combine):

- Information is complete and accurate but not timely (delayed / obsolete),
- Information is complete and current but not accurate,
- Information is accurate and current but not complete,
- Accurate, current and complete information is available somewhere in the Project but not transmitted or only transmitted with distortion to the relevant compiler.

Schedule update failure root causes

Quite logically, the root causes of these failure modes are the following:

- There is a (systematic) delay in the information collection process,
- There is a (systematic) bias in the information measurement process,
- The information measurement process does not cover all the relevant activities or is not relevant to actual physical progress,
- Information does not flow to the planning team.

Some of these root causes might overlap on part of the Project activities. Common examples of these root causes are:

- The last information available on a contractor's progress is often their progress as it was 2 months ago (due to poor expediting or several layers of subcontracting and reporting processing and consolidation),
- When documentation progress measurement is based on the engineer's declaration they tend to be systematically optimistic about the completion status of their documents; alternatively people don't want to hurt the Project Manager and do not report the actual status of the Project versus the plan (very common historical issue in centrally planned communist countries, also prevalent in countries with high respect for hierarchy),
- The progress reported is not really physical progress but cost progress, which does not give any real

indication of what has been done physically (common issue for fabrication and complex manufacturing),

- The Project schedule is so complicated and large that it is physically impossible for the planning team to review progress and forecast for all activities for each update cycle,
- Planners/ cost controllers are not proactive in ensuring that they capture all available information at its (reliable) source and cross-check it from a second independent source whenever possible.

Implementing a robust and consistent progress measurement system from the onset of Project execution is important. The frame on the next page lists particular areas of vigilance regarding progress measurement systems.

An Ineffective Information Collection Process: the Main Failure Cause for Schedule Updates of Large Projects

While delays in the progress information and bias in the measurement process are classically addressed by extrapolation and reliance on actual physical progress measures, the main stumbling block for Large, Complex Projects seems to be the information collection process, with a classical double whammy:

- Too large and complicated schedule that is painful to update,
- As a consequence, not enough time to seek and check the most difficult information at its source – namely, the people in the team in charge of this particular activity.

There have been several cases when upon intervening on a Project we have found substantial differences between the team informal knowledge (as retrieved by interviewing Project team members) and the formal representation of the Project status and forecast in the schedule.

Best practices in Progress Measurement Systems

Progress measurement needs to be consistent and adequately reflect physical progress. It is better to be conservative than to over-recognize progress. The following recommendations apply for various stages of project execution.

- Engineering: one issue here is that Contractors often promote aggressive progress recognition methods in particular because progress is often a basis for billing.
  - minimize progress recognition for starting a document,
  - rely as little as possible on engineers' judgment for document draft progress, rely rather on actual stages such as first draft for internal review,
  - make sure that production of a first version for external review does not exceed 60% progress,
  - and do not recognize 100% progress for the first Approved for Construction (AFC) version as there are often later updates (e.g. a first AFC version of the document can be weighted 90% only).
- Procurement & Fabrication
  - Make sure to rely on actual physical progress, require and approve a detailed schedule and progress measurement system from the suppliers / fabricators and roll-up the result at the level required.
- Service contracts and logistics
  - Only develop progress measurement if the services are not directly pegged to construction (otherwise, simply increase the weight of construction to account for the other related activities)
  - Progress is generally more difficult to assess. Determine a very limited number of physical quantities that can be used as trackers for physical progress.
- Construction
  - Make sure to rely on physical progress. Manhours spent is an indication of cost, not of progress,
  - Do not overcomplicate and devise a limited number of key quantity indicators (e.g. volume of concrete, weight of re-bar, length of pipes, number of spools etc.) that will be representative of physical progress and serve as a check to possible more complicated measurement systems based on detailed schedules.

Finally, do not update the weighting of each activity during project execution, because it would made comparison with previous progress measurements impossible. The imprecision created is generally negligible at the aggregated level and does not impact accuracy.

Faced by a tremendous update task, under time pressure, people will rather work on the easy (update from available systems) rather than the more difficult (understand actual progress from interviews and site appraisals). And there come situations where the really valuable information is not sought and what the schedule of the Project reflects has got only a distant relationship with the reality in the field.

What Needs to be Done to Have a Reliable Project Snapshot

It is not sustainable to drive a Project with an inaccurate schedule or cost model. Project managers need to be particularly watchful to avoid this happening. Here are some guidelines:

- Limit the complication and size of Project schedules to keep them manageable (think how many activities a planner can reasonably update in a day if the full work of listening to the people in charge is effectively done) (ref. Chapters 2 and 4 on the discussion on the size of the Integrated Project Schedule),
- Make sure that all relevant parts of the Project team are actively involved in the update sessions even by remote teleconference or other ways and fully understand the importance of the update process,
- Ensure that there are real conversations ongoing on the actual status of activities, and not just updates of documents or tables. The Project Control team needs to get a feel of the situation. This might require the involvement of the Project Control Manager to help having candid conversations about the actual condition of segments of the Project, and even travels on site to gauge the actual progress with an independent eye,
- Make sure that physical progress is actually measured and reported and not just busy-ness.

Re-sequencing versus Re-baselining

After a proper schedule update taking into account effectively the knowledge of the entire Project team, it might happen that the Project Completion Date appears to have shifted.

Should the Completion Date remain consistently delayed over few schedule updates (say 3 successive periods), the planner will then consider a new fundamental step in the schedule update: re-sequencing. Re-sequencing is different from re-baselining in that it is an optimization of the current schedule with the aim to protect the Critical Path, and it is not a full review and re-engineering of the full Project schedule and execution strategy.

Re-sequencing consists of reviewing the Critical Path and near-Critical Paths, the sequence and duration of their activities and identifying what optimization changes to the execution, such as doing some activities in parallel, increasing resources to reduce a task duration, changing the work sequence or method etc. are required to revert back to the original Completion Date.

Should any such changes to the execution be required, it is essential to discuss with the concerned parties (Engineering, Procurement, Construction and Commissioning) and to get their commitment prior to implementing those changes in the schedule, in particular when they significantly change the expected dates and resource commitment levels. This should happen through a formal comprehensive Management of Change process.

Such changes shall also be highlighted and explained in the schedule update narrative to keep everyone's confidence in the integrity of the schedule.

The narrative shall explain all changes made, including change to the logic, change of duration, in particular for activities not yet started etc. It is an essential record to maintain for contractual discussions (refer to Chapter 12).

Only if there starts to be a very substantial slippage of the Completion Date or if events require a deep rethink of the execution logic should the Project engage in a full re-baselining. Re-baselining is a very heavy exercise that impacts the entire Project execution (including many other documents and plans) and should be apprehended with caution (ref. Chapter 4). It generally requires approval by the Client and needs to be approved by senior management.

Conclusion: Listen to Your People – Use the Schedule to Transform Informal Knowledge into Formal Knowledge

Being able to transform the informal knowledge spread inside the Project team is an essential process capability for successful Projects, in particular when they are Large and Complex, and even more when the Project team is spread between different sites. The successful Project team will make sure it works perfectly – and will stress-test it from time to time. A sufficiently representative Integrated Project Schedule that is at the same time not overly complicated is essential; as well as the capability of the Project Control organization to seek people, get them to talk, and listen to them.

Reality is sometimes uneasy; but it is much better to know the reality, and structure this knowledge so as to be able to act on it. Too many Projects maintain illusions too long in their schedules, falling prey to a delusory perception of reality which will catch up sooner or later. Don't let this happen to you and be constantly ready to challenge an update to make sure that your vision of reality is not impaired.

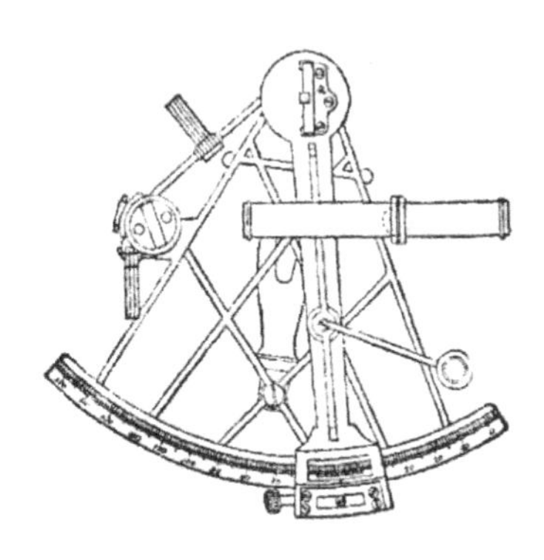

Chapter 9: Forecasting and Identifying the Project Drivers

Chapter Key Points:

- Contrary to Updating, Forecasting is not necessarily a bottom-up exercise. It requires to understand the underlying schedule drivers.
- The basis of Forecasting is the determination of productivity factors and/or proper communication with external parties.
- Schedule Forecasting approaches depend on the type of activity.
- 'Virtual float' created everywhere when there is delay in part of the schedule needs to be managed to avoid propagating poor productivity throughout the entire Project.
- In addition to conventional forecasting, float monitoring is a key tool, together with Convergence Planning, to identify those issues that could bring the Project to a halt if they were missing at the time they are needed.

Why Forecast our Future Trajectory?

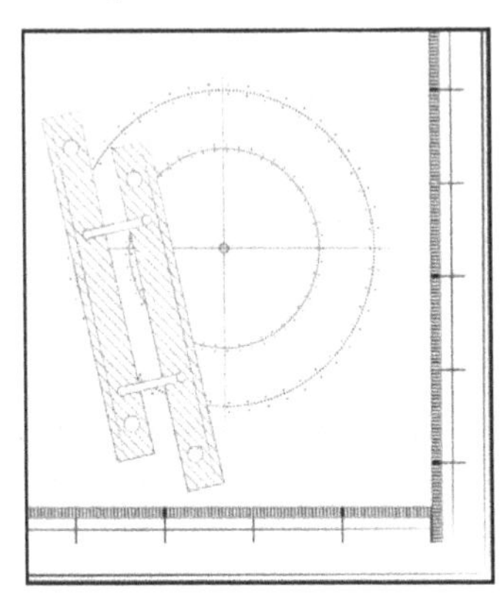

The previous Chapters were first about producing a useful set of maps for our expedition – with the right scale, level of details and consistency. Then we investigated what needs to be done to get an accurate status, the Schedule Update, which is the equivalent of taking a sight to determine position. In this Chapter, we will deal with Forecasting.

This particular process responds to the second key navigation question:

- Where are we going? (if we continue to follow the same trends)

It is an essential question that is a prerequisite to decision-making. Decisions will be about changing parameters, such as the speed, heading and the sails' settings, to reach destination if the forecast shows that we are deviating significantly from our end point.

The Three Steps of Re-Forecasting

Re-forecasting a schedule occurs in three steps:

1. Based on the actual progress of the **currently ongoing activities**, determine their expected re-forecast completion time;
2. Observe how the **Project activity network evolves naturally** as a consequence of the re-forecasting of the ongoing activities, and whether issues need to be tackled with regard to certain requirements;
3. Reforecast **future activities' duration and effort** (activities that are not yet started) based on the new knowledge available from ongoing activities or any new information that needs to be taken into account.

The first and the third component require activity re-forecasting methods (duration and possibly, effort level). The second component is the natural consequence of a well-designed and connected activity network with a minimum of artificial constraints. The latter will normally be ascertained as part of the quality of the Integrated Project Schedule, so that we won't elaborate further in this Chapter. We will thus concentrate on the issue of re-forecasting currently ongoing and future activities.

It is important for Project Managers to be aware that while planners quite naturally tend to reforecast activities that are currently ongoing, they might lose sight that new information is now available that can also be used to update certain future activities. It is often up to the Project Management Team to challenge the planners and ensure that the knowledge is effectively applied to non-started future activities, whenever appropriate.

Finally, when it comes to the Project completion date, in the forecasting process, a particular focus should be put on the activities on the Critical Path or on activities which are close to being critical.

At What Level is Re-Forecasting Best Approached?

Schedule Forecasting is Not a Bottom-Up Exercise

Contrary to schedule update, it is not necessarily appropriate to approach re-forecasting from a bottom-up perspective.

Bottom-up re-forecasting often leads to unexpected and overlooked consequences on the overall schedule; critical convergence points can make sudden and unexplained shifts to the right. Forecasting at a higher level will allow re-sequencing and rescheduling of future activities to accommodate detailed level changes: there often exist alternative sequences of work that can be implemented if the base plan is affected by early or late progress. The Project team should constantly evaluate alternative detailed sequences that can be implemented to take advantage of opportunities or mitigate effect of risks. Such alternative schedules/ sequences should be considered at all levels of the Project. For example when there is a delay on a critical valve in a module, alternative fabrication sequences to accommodate late delivery of the valve will be assessed by the construction lead, whereas the Project Management Team will assess alternative installation sequences of the fabricated modules considering various module delivery dates.

Bottom-up re-forecasting activity by activity will also not be meaningful if one does not understand the reasons for the deviations from the baseline durations, or for delays due to some critical activity. Typically, for example, delays in engineering could be either a one-off issue on a particularly difficult engineering piece of work; or a common cause issue running throughout the engineering activities; or delays due to waiting for some key data. An analysis of the cause of variance is needed prior to proceeding to a schedule re-forecasting, that will inform the reforecasting method. Many organizations fail on this matter due to the relative isolation from the Budget Owners and lack of business understanding of the planners. Then they rely on bottom-up reforecasting that have limited meaning.

Schedule Forecasting is also Not a High-Level Exercise

Project forecasting can neither be done at a too high level for the following reasons:

- The linkages between activities might create significant bottlenecks or unexpected consequential effects between functions, sites and unrelated activities,
- A subcritical chain of activities might suddenly become critical and drive the Project unexpectedly.

These effects would not be visible at too high a level.

In particular, while overall S-curves (generally presented by function or area) are a useful summary of schedules, they should be used with caution to re-forecast based on the current condition. The same holds for project-level Earned Schedule application. These representations aggregate too many factors; they are useful for high level reporting, but are much too simplified and often ineffective when it comes to forecasting. Actually, they would be a valid tool only if delays were only due to productivity without any critical chain or convergence issue, which is rarely the main factor for delays in Large, Complex Projects.

Alas, much too often, senior management does not want to deal with the complications of Integrated Project Schedules and draws conclusions from the observations of very high level S-curve trends. It is a very dangerous practice that can best give orders of magnitude, and worst case, lead to serious misperceptions – and under evaluation of the actual trajectory of the Project.

As a general observation, proper forecasting requires a birds-eye view of the Project execution, at the right altitude. It requires understanding fully what is the activity or constraint that actually drives the Project schedule, at least for that particular phase of the work. Forecasting requires a deep analysis and understanding of the drivers of the Project and it is what makes it a valuable art.

Basics of Schedule Re-Forecasting

Basic Re-Forecasting Approach

In its simplest form, schedule forecasting for a given activity or relatively linear group of activities of similar content will be based mainly on the comparison between the baseline and the actual progress, at some relevant level. The difference between the two will give an indication of the actual *schedule productivity* of the resources dedicated to this activity.

Using Earned Schedule terminology (ref. Appendix 6), schedule productivity, also called Schedule Performance Index (SPI(t)) is the ratio between the 'Earned Schedule' ES and the 'Actual Time' AT:

$$Schedule\ Performance\ or\ SPI(t) = \frac{ES}{AT}$$

The updated forecast for the duration of the activity is then the baseline duration, divided by the schedule productivity.

This simple approach can be further refined to account for specific effects:

- Learning curve effects – on some activities, it can be expected that the progress at the start of the activity will be slower, with an acceleration when the resources get used to the tasks at hand. Because this is a non-linear effect, the instantaneous productivity must always be considered with caution when used as a basis for reforecasting. Learning curve effects can be very impressive when more than 3 complicated similar operations need to be performed; the last operation can sometimes last less than half the duration of the first operation,
- Completion delay effect – on the other hand it is often very difficult to gain the last few percentages of progress, and the activity finalization forecast often

must include specific degradation factors for activity close-out,

- Resource level effect – a more refined productivity measure is per key resource, so as to account in the forecast for increasing or diminishing resources. However, caution must be observed with the fact that the law of diminishing returns applies to additional resources and their effect can be less than a linear extrapolation of productivity,
- Multitasking has an effect on productivity as well as more generally, change from one activity to another, or from one configuration to the other for the same resource. This leads to loss of efficiencies,
- Similarly, the Resource bottleneck effect – in the specific cases where the activity's actual progress can be related to a bottleneck involving one particular resource (personnel or equipment), this bottleneck and its forecast variation can be used as a basis for productivity estimates.

Also, this basic approach only works with the condition that there are no external influences on the execution of the activity – such as temporary resource unavailability, or interference with other activities from a resource or execution constraint. These pre-requisites need to be checked.

Earned Schedule can also be applied to an aggregation of activities with the condition that each activity's weight is available (it is generally proportional to its budget cost). Earned Schedule can then be calculated for all the activities together. For example, it can be calculated on a sample of activities such as the project's Critical Path. However, this remains sensible only in the case of activities of similar nature with little interference from the rest of the Project. It is hazardous to use this method on an entire complex Project or sub-Project.

Rule of thumb and best practice: *It is important to reforecast currently ongoing activities or meaningful Project phases after about 20-30% of progress. Earlier will not give very reliable insights; later... is too late for action. Thus whenever consistent groups of activities reach 20% to 25% cost progress, the schedule department should implement a specific review to reforecast the activity.*

Re-Forecasting Future Activities

The previous paragraph dealt with activities that are being currently performed. It is also important to re-forecast future activities in the light of available information, even if they have not started yet. It is the case, in particular, if present or past activities of similar nature give reliable schedule productivity data in the particular context of the Project.

It is important not to lose sight that new information can also be available from other sources that can also be used to update certain future activities. Such benchmarking information can be available e.g. from other projects currently operating in similar conditions.

This is particularly important for durations and impacts that will be recurring on the Project, such as climatic and seasonal impacts, infrastructure availability constraints (port, access road, heavy weight equipment), turnaround times for trucks or dredges to the load-out or dump area, etc.

The Importance of a Stable and Consistent Baseline

The basic principles of re-forecasting exposed above only work if the baseline is stable and its assumptions clearly available.

As explained in Chapter 3, it is essential that schedule rebaselining remains a rare event – because it is extremely difficult to reach a moving target. It is even more important to ensure that the process of reforecasting is done on a consistent baseline basis. Productivity must be measured with respect to a reference that is to remain stable. In case of rebaselining that changes the durations of activities, it is essential to update the productivity references as required so to ensure consistency and traceability of the productivity data from the first Project baseline.

Conditions for Forecast Quality

There are a few additional conditions that need to be met; otherwise the forecast cannot be of the expected quality.

Forecasts by their nature are not 100% accurate, if we could predict exactly what would be the consequences of the variances that are observed, then scheduling would not be as difficult and require the skilled people that it does. Forecasts need to be challenged continuously, every time new information or data is available to the Project.

Committed Budget Owners

An essential prerequisite is of course committed and involved managers that take responsibility for their share of the Project activities. In the Project organization, these managers are the Budget Owners that are ultimately accountable for the forecast of their Work Packages both in terms of cost and schedule, while the planner provides data, support and challenge.

It is important to avoid the syndrome of "padding" by Budget Owners, i.e. adding duration to cover unfortunate circumstances. Risks need to be dealt with at the contingency level and it is not adequate to have each activity duration padded by budget owners that would be anxious of being judged on their performance. This can happen at the time of the baseline estimate, but also in the course of re-forecasting.

Adequate Communication between Parties

Forecasts are only as good as the quality of the information that support them. It is essential to have an excellent flow of information between the different parties involved in the Project.

Best Practice: *A good planner does not remain at his/her desk. An important quality for a planner is the ability to communicate with all the Project personnel, to listen and identify any hint of delays or schedule variances, and to get people to talk about the progress of their work. Conversely, Project team members must be conversant with the schedule and inform the planners in case of changes of forecast.*

It is essential that all relevant information flows immediately to the Project Control team to be included in the schedule, cost and risk models of the Project so that their impact on the forecasted outcome can be established and analysed. Planners bear a large responsibility to get the right information from busy Project team members. Communication skills – and in particular, listening skills – are key competencies of a planner.

Cross Checking of Information

Irrespective of the level of details of the schedule, it remains a proxy of reality. It may be wrong as a model of that reality. Reporting may then fail to identify issues early. To mitigate that risk, it is important to cross reports & forecasts: example reforecast resource intensive activities based on productivity and check with suppliers' management their vision of the delivery date. This requires Budget Owners to be involved and also the planner to check its sources. Like in journalism, the rule of having two independent sources should be used on all areas where forecasting might be at doubt. This is particularly the case for suppliers of long lead items that can't easily be substituted, fabrication contractors and other contractors which need to implement activities that are critical for the Project schedule.

Basic Forecasting Approaches by Type of Activities

Best Practice: *Forecasting requires the planner to have an understanding of the business. This is why in large Projects with a team of several planners, it is often more efficient to specialise certain planners to the maintenance of specific Detailed Functional Schedules. However, it is important that enough planning resources are retained to maintain appropriate visibility and update the E-P-C chains in the Integrated Project Schedule.*

In the following sections we explain the best practices by type of activity. However, looking carefully at the interfaces between functions will often lead to the discovery that the issue is not the intrinsic productivity of the function, but a poor management of interfaces, and subsequent delays at those interfaces, which might not even be related to the intrinsic productivity of either function.

Actually what is really most important is to identify the chain of activities that drives the Project at any point in time. While in the subsequent subsections we explain good practices by type of activities, this higher concern needs to remain present at all times during the update of the schedule.

Engineering

Beyond the Project start-up activities which are deliverables/ milestone driven (ref. Chapter 5), engineering is purely personnel activity (sometimes mobilised through a service contract), and the deliverables to be produced are documents or other similar deliverables. Work is often split between CTR (Cost-Time-Resource) elements which are Work Packages for engineering.

For engineering, Earned Schedule can be a powerful forecasting tool provided there is a sound progress measurement basis.

The Integrated Project Schedule should not contain all the detail of the CTR and engineering deliverables – it is the role of the detailed engineering schedule (and of the document register which also contains planned and forecast dates). The Integrated Project Schedule should contain a subset of engineering deliverables that are critical for Procurement and Construction, which generally are deliverables resulting from a chain of engineering activities. Vendor data required to complete Engineering also needs to be identified clearly. Thus, it is not relevant to measure schedule productivity on the activities of the Integrated Project Schedule: productivity should be measured on the Engineering Detailed Schedule or Master Document Register to anticipate when the critical deliverables will be available. This is applicable both to internal and external engineering activities.

The last 5% to 10% of engineering physical progress are generally much longer than envisaged by most plans, thus forecasts need to be prudent, until the actual effort for that phase has been ascertained for the particular context of the Project. However depending on the situation it might not impede linked activities such as procurement or construction to proceed, with an associated risk of rework and poor quality. This needs to be carefully analysed when establishing the schedule.

Unless supported by actual data (historical or benchmark), estimated engineering durations tend to be optimistic. Remain prudent on schedule re-forecasts that would shorten the engineering phase and avoid significantly shortening forecasts before the full cycle of documents have actually been achieved (including progress of some documents up to Approved for Construction status) to have benchmarks applicable to the particular Project's situation (allow at least 30-40% engineering progress on those first sets of activities to measure actual schedule productivity).

Procurement

Procurement refers here to the purchase of items. The following section will deal with service contracts (purchase of services).

The pre-award phases and the first post-award phases (up to the kick-off meeting with the supplier, or even sometimes up to the start of manufacturing if there is a heavy part of pre-production documentation production) are relatively under the control and the view of the Project team. They need to be forecast based on milestones and generally cannot be related to a productivity measurement.

Observation of the Project team's performance until contract award and effective kick-off is a useful set of data for updating the forecast of similar activities in the future (as it might depend on stakeholders' involvement and behaviour particular to the Project, such as senior management and end-Client's approval cycles). Often, award approval requirements will be clarified only after the start of Project execution and may lead to significant changes in the lead times that were assumed in the baseline. This can imply substantial changes in the Project execution strategy, beyond re-sequencing of activities.

Then, the manufacturing phase until delivery is generally hidden from the Project team and only accessible through intermediaries:

- Supplier updated detailed schedule,
- Expediter's follow-up dashboard (which may include data extracted from Quality Control inspections and other sources such as visits).

It is important that the Project's planners be involved in specifying the level of detail and the interface milestones that will be required from the supplier in its periodic manufacturing report. It is also important to ascertain that the supplier's physical progress is effectively measured against clear physical deliverables. This is generally straightforward in manufacturing activities. For remote suppliers reality checks are essential, e.g. by requiring pictures to be inserted in inspection reports.

Due to the delays in establishing and receiving those reports, the updates from the supplier might not be the freshest and they need to be checked against the latest knowledge of the Budget Owner and of the expediter. It then needs to be extrapolated in terms of delivery forecast, based on a discussion with the expediter and the supplier.

Contracts for Services and Logistics

Contrary to the field of Cost Control, contracts for services[1] and logistics are often relatively easy to follow and forecast in schedules.

From the schedule perspective, there are grossly two types of Contracts for services:

- Contracts for services that will support major construction activities with minor impact on their duration, such as Quality Control/ quality inspection, construction support activities, etc. These activities are rarely identified as such in the schedule because the follow-up of construction activities implicitly track the performance of the work by the contractors,
- Contracts for services that can have a direct impact on the duration of the work, such as pre-construction process qualification, rental of key construction equipment, availability of concrete, logistics etc.

In the particular case of logistics, standard timing for the performance of the transportation of items between two points can often be derived from the observation of the first actual transportations (taking into account all the hurdles such as customs clearance). These durations then must be spread throughout the schedule re-forecast in particular when logistics form an integral part of the Critical Path.

In addition to schedule productivity, schedule forecast must consider potential cases where construction activities are held up by late arrival of material or equipment, which can in certain cases be the driver of the schedule.

---

[1] Often called "Subcontracts" in Contractor organizations

Fabrication (in fabrication yards)

Depending on the contractual setup, fabrication can be either similar to procurement (generally, for straightforward off-the-shelf items) or to construction (for complicated custom fabrications). Refer to the relevant section.

For complicated fabrications the roles of the Project site representative and of the Budget Owner are essential in ensuring that an accurate report on physical progress is available. First, some work is generally required to determine appropriate physical progress measurements; for example, an approach by weight of steel welded might not be appropriate in all cases. A detailed weighted schedule might be a good reference, but needs to be reviewed with caution.

Fabrication yard schedules should be resourced so as to be able to observe and challenge the manpower curves, in particular with respect to worksite manpower density limits, and the overall capability of the yard. Depending on the type of fabrication and of the fabrication process specific trades might be critical and drive the schedule. They need to be identified as the relevant resources will play an important role in considering any acceleration plan.

Complicated fabrication schedules are often driven by the availability of detailed engineering drawings or procured items, and when establishing the fabrication schedule, particular care should be given to establishing the proper interfaces with the engineering and procurement activities, taking into account the necessary transportation and logistics durations. The quality of the expediting process and of the forecast dates for the arrival of the equipment is then essential to avoid last minute re-scheduling changes which necessarily hamper productivity and favour rework.

If the fabrication schedule is driven by the actual fabrication activities, then productivity factors can be derived to forecast the completion date (focused on the critical trades). This forecast can be adjusted based on expected changes in the allocated resources (being careful that two shifts do not mean double productivity, nor does increasing manpower if there are congestion issues). The forecast needs to be carefully reviewed in the case of yards working at the same time for several clients as manpower can be shifted to or from another clients' work.

In general, it is important that an appropriate allowance be available in the Integrated Project Schedule forecast (but not communicated to the fabrication contractor) for:

- Actual fabrication productivity
- Such events like engineering changes and delays in the deliveries of equipment depending on their historical frequency.

Construction (on dedicated site)

Site construction is generally composed of three main phases:

- Site preparation
- Facility construction
- Pre-commissioning and commissioning

It is essential to understand what the activities that will drive the construction schedule for each phase are, and that depends on the type of facilities. Only after that analysis can sound re-forecasts be issued.

Site Preparation

For onshore facilities, the first part of the construction schedule is always driven by earthworks (related to the volume of soil to be displaced) and ground preparation. Depending on the type of soil and the requirements in particular regarding seismic resistance, specific soil preparation might be needed that can be very long. For example, clay soils require preloading for a given duration that cannot be diminished easily; piling might be necessary on unstable soils or soils that could liquefy in case of seismic shock, etc. There is always a large uncertainty associated with the duration of soil preparation; however these activities generally require limited engineering and procurement and can be started during the time in which more elaborate engineering and procurement proceeds for the rest of the facility. Depending on the type of soil preparation, reforecasts need to be established based on observed productivities.

Facility Construction

Construction activities can often be properly reforecast using Earned Schedule and schedule productivity measurements.

The actual facility construction's driving factor will depend on the dominant structural element of the facility that needs to be installed on site and cannot be prefabricated. For spread-out, predominantly steelworks facilities such as most oil & gas processing facilities, piping erection is generally the driving activity. For other facilities that rely on large quantities of concrete (for example nuclear facilities in general), related activities of steel reinforcement installation and concrete production and placement will mostly drive the schedule.

A major issue at contract kick-off is to define appropriate physical progress measurements that are both representative and simple, and can be tracked effectively. The involvement of the Project Control team in general, and of the scheduling team in particular, is essential to make sure that physical progress is measured. The most important is to ensure that a measurement is found that is simple to track and report, and still remains representative of the progress. It is worth remembering that Man-hours are not physical progress, as 'busy-ness' does not necessary imply physical progress!

The Contractor is then required to report physical progress as per this mutually agreed measurement principle, after having produced a baseline based on its execution philosophy. In the field, this physical progress measurement must then be checked independently, at least by sampling (this is often the role of Quantity Surveyors). Productivity factors can then be derived that can be used as a basis for independent re-forecasting.

In addition, it is important to identify if the actual facility construction will be driven schedule-wise by inherent construction limitations or by interfaces with other activities such as engineering or regulatory compliance issues. Setting up the right interfaces with the rest of the Integrated Project Schedule is essential, together with a good analysis of the situation. For example, in the nuclear industry there have been many cases of construction being driven by the rate of engineering review by the regulatory authorities.

Once the driving activity for facility construction has been identified, either, productivity factors can be derived that allow proper re-forecasting; or, if driven by external events, productivity of these external events need to be observed and used as a basis for forecast.

Commissioning

Finally, pre-commissioning and commissioning activities (including pre-operational testing) is generally an activity that is underestimated in terms of complexity and duration. In addition, it is too often used as a buffer during facility construction: additional durations of construction get compensated by equivalent reduction of the commissioning duration! This inevitably leads to delays that are unfortunately revealed during the last months of construction. As a minimum, the initially estimated duration for commissioning needs to be maintained in the Project schedule as a health measure. During commissioning, productivity factors based on number of systems, number of tests or other appropriate measurements specific to the facility can be used to reforecast.

Because of the logic switch from geographical location to systems between integration and commissioning, and possible concurrent remaining construction activities it might be difficult to understand what is driving the schedule at any point in time. Close update and analysis of the detailed commissioning schedule needs to be maintained at all times.

Offshore Marine Construction Operations

Offshore construction will generally be separated in two different phases:

- Mobilization and transit
- Actual offshore construction work

It is relatively easy to derive schedule productivity factors for different types of offshore construction work.

Mobilization activities' forecast difficulty depends on the possible repetitiveness of the operation; they are in any case quite short so that it is generally not a major issue.

Like in other areas, care must be taken on the interfaces with logistics for the supply of material offshore, which may drive the schedule if improperly managed. The possible impact is often compounded by the remoteness of the site.

The Fallacy of 'Virtual Float'

In Project life, activities generally tend to shift to the right – to a later date. This actually will create float in all the activities which are not critical.

This has a detrimental effect because it means that most Project contributors will in effect be given more time to complete their tasks. The effects of this situation are well known:

- Because they are given more time, people will take more time to do their task (Parkinson's law);
- Or, they will start later which does not improve the odds of the task being finished on time (Student's effect),
- And, interfaces will be another excuse not to finalize work that could be finalized, thus, the disease will spread throughout the Project.

We call this situation creating 'virtual' float. It is a vicious circle. The more the Project is late, the more 'virtual' float is created, the more time and money will be spent to achieve the same results.

How can a Project overcome this vicious circle?

- The Convergence Plan if it is applied with discipline is an excellent way to maintain the pressure on all Project deliverables even if one sequence of activities faces a hurdle – because the dates at which the deliverables need to be achieved remain fixed,
- When re-forecasting the schedule,
    - instead of letting the entire network of activities spread in time, add "buffer" resources on all non-affected sequences of events (owned by

Project management) to force these sequences of activities to remain on the same overall schedule,

- alternatively, force a constraint on the finish date so that delayed activities show negative float and no 'virtual float' is created in the schedule network. This practice has the advantage to tag delaying activities with negative float, which gives a clearer delay analysis for Extension of Time claims. On the other hand it introduces constraints and negative float in the schedule, making it unrealistic overall so it needs to be used with prudence and only when required by contractual strategies.

We prefer the method where buffers are added to constraint non affected sequences of events. This essential practice is explained in the following figure.

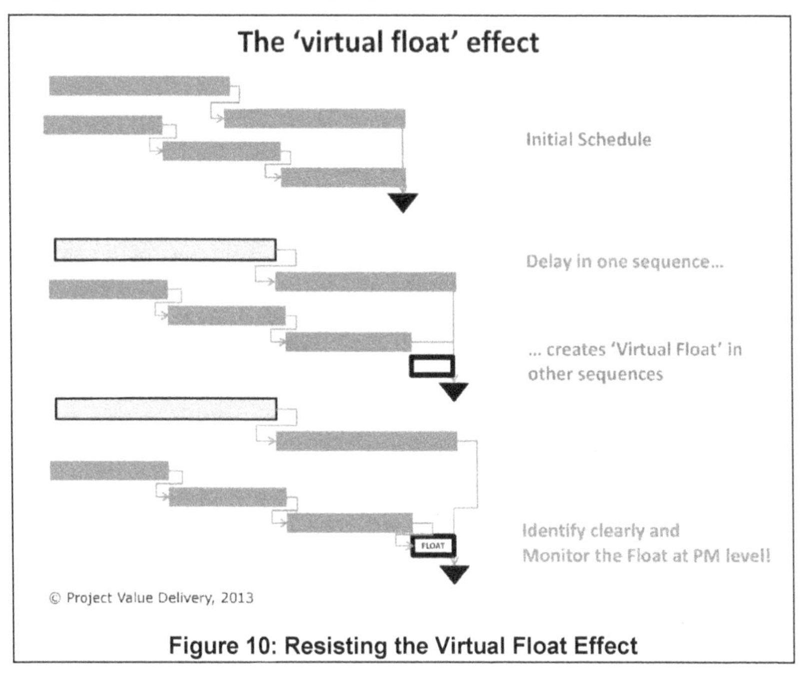

Figure 10: Resisting the Virtual Float Effect

Real World Project Re-Forecasting

In most cases (except frontier Projects using new technology), major Project delays are not due to improper estimates of pure productivity. In the real world, delays are due to interface problems – the small missing piece that makes the entire Project crawl to a halt.

These effects are generally not covered in re-forecasting processes, whereas they do in effect cause the largest disruptions of Project schedules. How can we capture these issues before they happen, act to avoid them, and if they are unavoidable, re-forecast for them?

Project Schedule, Critical Path and Critical Chain

Productivity-based reforecasting is fundamentally like re-forecasting the Critical Path of the Project. The Critical Path is the logically-linked sequence of activities that determines the duration of the Project. If any of these activities is longer by an amount of time, then the entire Project will be longer by the same amount.

A more advanced concept is called 'Critical Chain' and has been coined by Eliyahu Goldratt as part of his well-known Theory of Constraints. In this approach, a critical chain is determined through the Project taking into account in addition resource limitations and constraints. The Critical Chain can be different from a logically linked sequence of events because of specific resource constraints which might have an effect on different sequences of activities. Determining a Critical Chain requires a full resourcing of the Project schedule and is in fact rarely used for Large, Complex construction Projects.

In real life however, the nice linear view of Critical Path or Critical Chain is often contradicted by the fact that some sub-critical sequence of events suddenly becomes critical for the Project and unexpectedly drives the Project delivery. It can be due to some resource bottleneck and more often than not, by unexpected external events. This creates a discontinuity, a change of priorities and thus inevitably, if not properly anticipated, a high disturbance in the Project execution strategy.

There are ways to increase the robustness of the schedule at the Project planning stage to be less susceptible to this discontinuity effects and they have been explained in Chapter 7.

In this section we will concentrate on the problem of anticipating such discontinuities in criticality.

Convergence Planning is a tool that is typically designed to keep in check and identify such unexpected changes of priorities. If a Convergence Plan is well designed it should contain the critical deliverables that form:

- the end of significant logical sequences of work,
- or critical deliverables that are required for the rest of the Project

Proper Convergence Planning will indicate in advance whether these deliverables will be on time or not.

This information can be used to focus management attention on those parts of the Project where deviations happen, avoiding significant impact, and/or highlighting with a few weeks advance that a problem is developing.

Accounting for Changes

Although Projects generally try to minimize them, there are always changes to the scope of Projects. This will lead to the addition or removal of activities in the Integrated Project Schedule. Unless changes are so significant that a rebaseline is necessary, those changes should remain minimal. Forecasting can thus normally neglect them, unless they impact directly the Critical Path.

Float Monitoring Techniques

On top of a basic usage of convergence planning, float monitoring is a very powerful technique that can anticipate with more advance notice future deviations in key deliverables. What is difficult is to identify which are those deliverables that would need to be tracked. As a first approach, the deliverables identified in the Convergence Plan are excellent candidates. They can be complemented with other deliverables that are considered particularly key for the Project.

A word of caution however: float monitoring only works properly if the Integrated Project Schedule updating process works properly, and if the forecasting process is implemented with at least productivity-based re-forecasting. The recommendations developed about building a quality Integrated Project Schedule must have been followed properly.

There are two ways of monitoring float:

- Monitoring activities' floats as calculated by the Integrated Project Schedule program, either:
    - Total Finish Float as an adequate measure of how much float is available before impacting the Project completion date; or,
    - Free Finish Float to measure the time available before the next activity;
- Monitoring the available float with respect to a fixed date, e.g. a convergence plan gate or any other meaningful signpost.

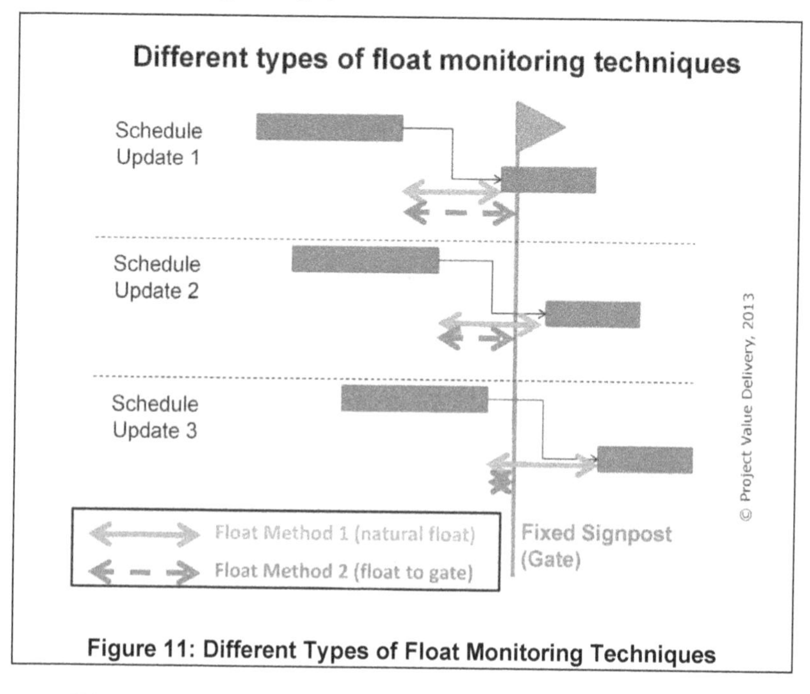

Figure 11: Different Types of Float Monitoring Techniques

The advantage of monitoring float as it evolves with the schedule (first method) is that it gives a good reflection of the actual available time; however it does not oppose the fallacy of 'virtual float': as activities (generally) move to the right, more time will be available and this is not a good practice to leave additional time to those streams that are not driving the schedule (see section on virtual float above). This is why we will prefer the second method where float is measured against a fixed date, which can be conveniently set as being a convergence plan gate date.

The deliverables that have been identified are associated with the finish date of the relevant activities in the Integrated Project Schedule. The finish date of these activities can be downloaded at the end of each schedule update and re-forecasting cycle, and compared to the required delivery date (for example in Excel). The difference between the latest forecast of the finish date and the required delivery date is the current float for that particular deliverable. In this method, the required delivery date is fixed and not allowed to vary.

Tracking the float and how it evolves from one Integrated Project Schedule update to the next will give very quickly, after 3 or 4 update periods, a very good idea of the trend. It is amazingly easy to anticipate in advance whether the trend is that the deliverable will be on time or not, as shown in the following schematics:

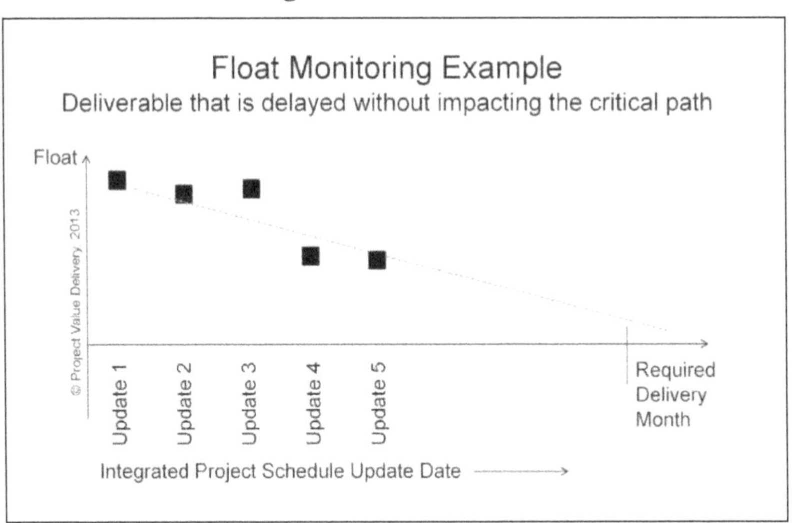

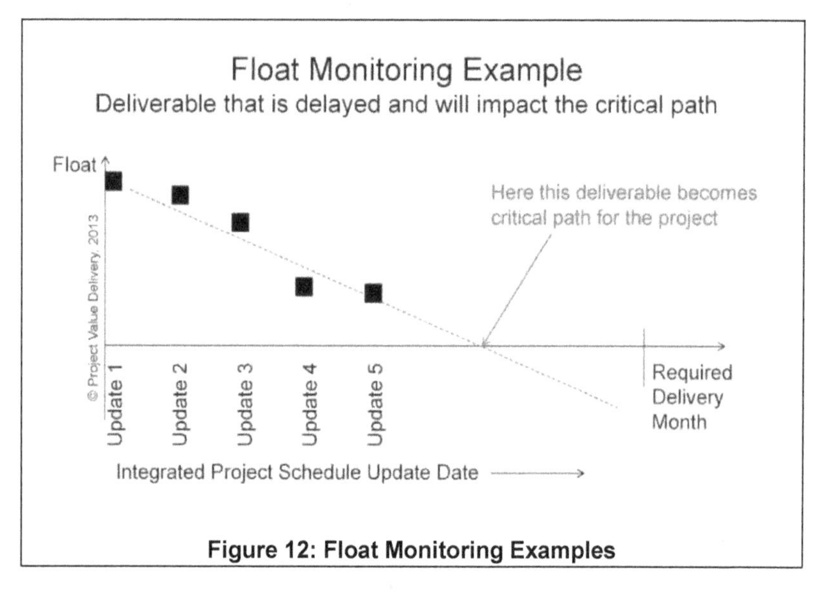

Figure 12: Float Monitoring Examples

These individual deliverables' diagrams can be combined in a high level dashboard which identifies clearly what are the trends that do not evolve properly and require investigation and possible action:

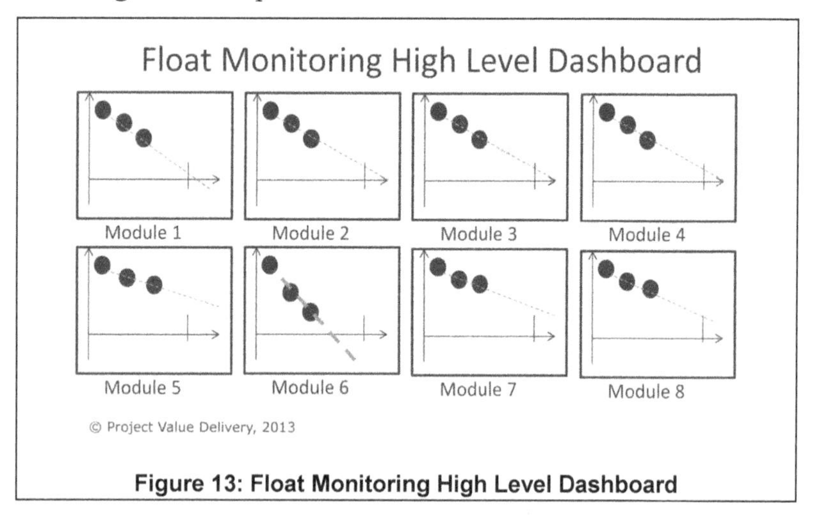

Figure 13: Float Monitoring High Level Dashboard

This amazingly simple method is extremely powerful and allows much earlier anticipation of issues in Projects. Its power is that it uses available information in a visual way. It is rare on Projects that the evolution of activities from one update of the Integrated Project Schedule to the next is considered and this method does make use of this available but unused information. It is a kind of ecological re-use of available information painfully created by the team and not really used to its maximum.

Float monitoring leads us to quote one of our Project Value Delivery most favourite principles, which unfortunately has to be highlighted to too many Project teams during Project reviews:

> **"An activity that is delayed by one month every month will never happen."**
>
> **(Project Value Delivery)**

How many activities in your schedule re-forecast get simply pushed by another month every month?

Project Delay Heuristics

To finish, a very simple heuristic is available that stems from the 'black swan' long tail mathematics, publicized by Nicholas Taleb. It is simple, yet ferocious in its effectiveness - even if only a few Project managers do believe in it when applied, it does work!

It simply states that if you are late by T compared to your baseline schedule, even after a deep review, chances are that you are ultimately going to be late by at least 2 x T.

You are warned (as well as your financing people)!

Conclusion

Schedule reforecasting is difficult because it takes a good understanding of the business. Yet Project teams do not spend enough time supporting their planning teams in that task.

Beyond conventional and easy productivity-based re-forecasting – that needs to happen at the relevant level to be effective – what is to be feared are discontinuities in criticality due to the unexpected change in Critical Path/ chain for the Project.

Simple albeit exceptionally powerful tools have been described in this Chapter to anticipate these unfortunate events. They draw heavily on the concept of the Convergence Plan and the identification of those critical deliverables for the Project.

One of the keys is to have the discipline not to take advantage of the additional artificial float produced by an overall Project schedule that would slip to the right, but to always compare Project performance to its baseline. This brings discipline and a stable target. The Convergence Plan is a key tool to maintain that discipline.

Spend the time on schedule reforecasting to understand what actually drives your Project. Once you'll have understood it, and if you can anticipate any change in this constraint, you'll be on your way to Project success. Because then you will take the right decisions.

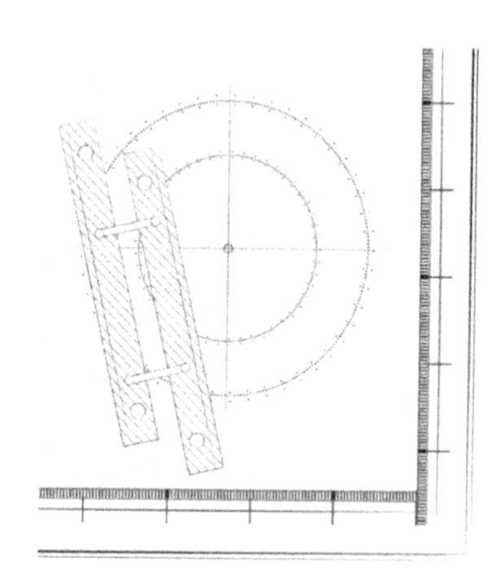

Chapter 10: Managing Changes: the Need for Schedule Agility

"Obstacles are those frightful things you see when you take your eyes off your goal" - Henry Ford

Chapter Key Points:

- No plan will survive the first encounter with the bulk of reality.
- Clear Project objectives and purpose are essential for project success and to respond to changes.
- Your schedule needs to be agile enough to cater for the events of Project execution.
- Avoid intermediate constraints as they will create significant inefficiencies.

Introduction: Your Plan won't Survive the First Encounter with Reality

What is the Integrated Project Schedule for? General Eisenhower used to say: "*In preparing for battle, I have always found that plans are useless but planning is indispensable*".

There is one thing we can predict with certainty: your Project execution is not going to happen as per the plan. And the larger and more complex the Project is, the less its execution will look like the baseline plan. Unpredictable things will happen, the Owner/ Client and/or stakeholders will change their mind (several times), and more...!

It is still important to go through the planning process so that you can understand the drivers of the events that will unfold, the depth of your actual strength and the general parameters of the environment where action will happen.

Another quote is famous in the military (where by definition complex Projects are being executed in highly volatile and unpredictable environments, the 'fog of war'). It comes from General von Moltke, a famous Prussian strategist from the second half of the 19th century (and head of the Prussian General Staff for 30 years): "*No plan of operations extends with certainty beyond the first encounter with the enemy's main strength*". Let us reword this for Project execution:

> **"No Project plan will survive the first encounter with the bulk of reality."**
>
> **(Project Value Delivery)**

This being posed as an eternal principle of Project management, what do we need to do to avoid losing control when we encounter the bulk of reality? The secret is to maintain flexibility.

The Importance of Clear Project Objectives

Our experience is that Project Managers are rarely entirely clear about their Project objectives, beyond the boilerplate expectation to deliver "on time, as per the requested scope and on budget".

Project objectives are a set of statements that describe in a nutshell, what the purpose of the project is and what are the priority areas that will be considered when evaluating the performance of the Project execution.

Staying only on the topic of money, what are the expectations in terms of cash flow? Acceptance of risk for possible additional revenue? Changes in the spending profile over time? Financial performance at reporting period end? How do all these expectations fit together? Are they consistent? Which ones are the most important, priority ones, and which ones can take a back seat? How do all the requirements fit into the 'Cost, Schedule, Scope' triangle?

The Project Manager, at the onset of the project, should have a clear roadmap of what are the main objectives of the Project, and what are less important issues. This roadmap should be explained in 10-15 statements that define what exactly is being expected from the Project.

It is not because the Project is to execute a very detailed contract, which is supposed to reflect the expectations of the client, that Project objectives are not important: there are many other stakeholders, and even the Owner's expectations often need to be expressed more precisely.

Indeed, Project objectives need to be defined together with the main stakeholders of the project at its onset. This worthwhile exercise will allow the Project Manager to uncover potentially hidden issues and motivations.

Unfortunately a proper Project objectives setting process is often skipped at the start of Project execution because of the pressure of operational issues. It is essential that Project Managers take the time to determine these Projects objectives in detail, in particular because they will inform what response to make to inevitable changes during Project execution. Refer to our *Project Control Manager handbook* for details about the recommended Project start-up process.

Stop Taking Change as an Exception: The Need for Agile Project Planning

In Project delivery, in particular for complex Projects, internal or external change is not an exception. It is rather the rule. Most of the Project management effort is actually spent in managing deviations and changes to the initial plan, with the aim to adapt to changing circumstances while disrupting the least possible the Project. How come, then, that traditional Project delivery processes are usually designed to manage change as an exception? Would it not be more effective and powerful to design Project delivery processes around change and agility as a core component, while preserving the effectiveness of the Project team?

It is certainly important to control changes to the scope to avoid scope creep, project disruption and Project contributor's fatigue; and this is one of the prime functions of the Project Manager. Still, keeping agility when it comes to the means and activities to reach that scope is an essential need – too often forgotten in bureaucratic Project-driven organizations. Quick decision-making is often the key to success, which requires a streamlined decision chain.

The Reality of Project delivery Success is About Agility

Most conventional Project management models and related processes are designed around a progressive, linear Project delivery, where processes are supposed to deliver predictable results, one after the other. Change is generally vilified as something to avoid – in particular to avoid the too frequent scope creep, to avoid rework and to maximize efficiency.

However, every Project practitioner knows that in real Projects, things never happen that way. Unplanned and unexpected events occur, changes are brought in by stakeholders, natural events or by the very characteristics of Project delivery, and iterations need to be done to come up with the right solution. Often the Project Manager himself introduces additional change on top of the change to mitigate the impact of an external change through re-sequencing of activities.

What makes the difference when it comes to Project success has often more to do with how agile the organization was to account for these changes and update its Project execution strategy and tactics; and not whether it forcefully managed to bring the situation back to the baseline plan.

In particular, the last steps to Project completion and the last contributors to Project delivery – construction and commissioning – are those that need to demonstrate the maximum agility to allow the Project to be delivered. They often have to re-plan continuously to take into account changes in delivery dates and readiness dates of different components of the Project. Experienced Project practitioners know that for construction and commissioning, flexibility and agility within a framework that guarantees control, quality and safety is the appropriate management response that successful Projects implement consistently.

Project Management Process Reloaded

In the IT industry, a set of Project management practices called Agile Project Management already accounts for the need to be very flexible when it comes to software development, with frequent reviews and updates of the planned activities, taking into account the discoveries done by the team during the Project delivery (software coding, in that case, which is quite a creative and difficult-to-predict endeavour). Decision-making happens quickly, through frequent reviews of the scope and progress. These reviews must be planned on a very regular basis (weekly).

Contrary to the common belief, Large Complex Projects need also to be agile – possibly with a lesser frequency than IT Projects because the end-objective might be better defined, but often with much larger possible impacts in terms of cost and schedule.

Why not, then, design the Project delivery process around agility rather than around trying to stick by all means to a set baseline?

Obviously the Project delivery process should not fall into the trap of an ever changing plan and uncontrolled creep in scope, time, quality or cost; and a baseline is always an important reference to come back to that enables performance comparison and measurement. Still, there are some ways to design Project delivery processes that take change, and subsequent agile re-sequencing, as a normal event in Project delivery.

Upending the Project Delivery Process

Let us consider now as a thought experiment that in fact, the main objective of Project management is actually to manage change and deviations to the plan. That establishing and executing the baseline plan is only a very small, easy part of Project management.

There is a threshold for changes and variances beyond which no particular action is required – it is when small variances happen that do not put in question the general logic of the execution plan. However should any deviation or change happen that steps beyond this threshold, immediate action ought to be taken. In this instance, reactivity is the most important factor to considered: the actual ripple effect of the event needs to be assessed quickly to understand whether it could be a threat to the Project delivery and how other activities would need to be rearranged. Two elements need to happen quickly:

- A clear decision taken by the Project Manager after due consideration of the consequences and the pros and cons in terms of disruption to the Project,
- An update of the plan for everybody on the Project to have an updated reference of the expected unfolding of events.

This requires that the Project management team has spare resources available to tackle this work; or, said differently, that it has managed to keep the work of normal administrative tasks at a controlled level that allows discretionary time to tackle those changes effectively.

Projects will fail if they:

- Have excessively complicated and developed schedules and cost models,
- Did not organize themselves in terms of processes, resources and systems to minimize the time spent on normal administrative tasks,
- Require the mobilization of external resources to manage changes.

These Projects will not be agile enough to decide and integrate change quickly, understand its effects down the line, and respond accordingly in a timely manner. Instead those Projects will just be in reactive mode to events that happen – they will have a direct reflex reaction without the proper thinking and strategizing. This might lead to an unrecoverable loss of control. Unfortunately, that is what is being observed too often!

Why You Need to Avoid As Much As Possible Intermediate Constraints in Project Execution

In Project execution, flexibility and agility are thus key success factor. Requiring the completion of intermediate deliverables adds constraints to execution, which can sometimes have significant negative consequences on Project success. Whether prescribing or executing, avoid as much as possible to introduce intermediate constraints. Because it is a very common issue, we devote an entire section to this problem.

In this section, the terminology 'intermediate constraints' is used to cover those activities or deliverables that are made mandatory in the Project execution by stakeholders. They are not the same as control gates such as the convergence monitoring gates, which are self-imposed intermediate check points for Project execution that serve to regulate the effort.

Intermediate Constraints Create Inefficiencies

It is a well-known fact from the Theory of Constraints applied to Project schedules that introducing fixed constraints for specific activities or deliverables in the middle of a chain of dependent events creates inefficiencies, in the form of additional waiting time for some resources. It does not allow the Project to benefit from the full potential averaging of the natural variation between the different activities.

In a more complex Project environment, intermediate constraints tend to create the same fundamental inefficiencies, related in particular to assignment of resources to complete the milestone in a way that might impact significantly other activities required for the Project. However it is not the major effect – the main issue is about making the schedule less flexible.

Common examples of such artificial constraints include:

- Intermediate contractual milestones for availability of supplied modules or equipment, while some flexibility could allow the works to proceed,

- A payment or contractual penalty milestone for 100% delivery of material diverting the focus from equipment or site mobilization whereas the full material delivery is not required to start the works,
- Certification processes that are not mandatory for the progress of the work, or are made mandatory too early.

Sometimes these constraints are self-imposed and more subtle, like for example awarding a single contractor for part of the works which are critical, thus requiring focus on feeding this operations with drawings and materials to ensure continuity. Splitting the contract in two, or requiring additional machines, might have diminished the disruption of the Project.

Intermediate Constraints are Mostly an Obstacle to Flexibility

In complex systems, a significant role of the Project team is to account for unforeseen events by playing around with the Project tasks by rescheduling, re-sequencing, and possibly changing the resources involved. Intermediate constraints and intermediate mandatory activities (as well as any resource constraint in time) add significant rigidity in how the Project schedule can be reshuffled in case it is needed.

The effect of such constraints in the Project execution schedule can be extremely significant. Intermediate constraints can be a significant aggravating factor to Projects facing unforeseen circumstances. Ultimately the delivery of the intermediate constraint might become temporarily the main driver of Project execution.

What Intermediate Constraints Could be Acceptable?

In certain circumstances, it may be unavoidable to specify intermediate constraints. So as to minimize their impact the following rules should be followed:

- Intermediate constraints should result from a stream of activities and resources as independent as possible from the rest of the Project;

- There should be a significant buffer between the planned availability of this constraint and its required date for the Project so as to minimize the possible impact of a delay on the rest of the Project execution.

What Should You do When there is an Embedded Intermediate Constraint?

The priority should be to avoid as much as possible intermediate constraint. Often, they are more warranted by some political issues than by real physical requirement.

Still, should it be unavoidable to have an intermediate constraint that is deeply embedded in the overall Project execution (for example due to a stakeholder requirement during the discussion of a contract or an unavoidable interface issue), the following general recommendations would apply:

- Try to keep as much flexibility as possible for the required completion of this deliverable (e.g. by implementing a window of availability and a notification sequence),
- Examine how the realization of this deliverable can be made as independent as possible from the rest of the Project execution in terms of resources and linkages; introduce a significant buffer for both the realization of the deliverable and its required date on the Project.

A common process for the first point is not to include penalties for delays on the intermediate constraint, or to include a clause that waives these penalties depending on the actual final completion date of the overall Project.

Control through Intermediate Constraints is a Waste

Sometimes, intermediate constraints are introduced by stakeholders as a way to control the Project. It might even be that decisions regarding the Project continuation be taken at these intermediate milestones. It turns out to be a very costly way of controlling the Project during the execution phase, because of the inflexibility it will bring to its execution. It will tend to make the Project execution a series of small Projects in series, from one control constraint to the next, and will tend to remove any gain from parallel execution with the final end in mind.

There are many other ways to effectively control the execution of a Project without creating the burden of intermediate physical constraints, and they should generally be preferred.

Conclusion – Be Flexible but Identify and Uphold the Project Purpose

Defining proper Project objectives are key to Project success. We need to go further here and speak of Project purpose – an inspiring purpose that mobilizes the Project team. The Project purpose needs to be clearly defined, clarified and communicated. All the actions taken to reach this purpose should subordinate to it.

The path that will be followed to achieve the purpose should not be as important as reaching it. Thus, while the Project plan might change depending on the circumstances, the Project purpose remains.

Therefore, instead of trying to maintain the plan fixed as much as possible, we now need to take the view that what should remain really intangible is the Project purpose, while the plan should be allowed to be modified by events.

Any externally imposed constraint on the execution plan such as intermediate fixed constraints are often the scourge of Project execution. They are often introduced by stakeholders or as a way to apparently control the Project. Make sure to avoid as much as possible this trap.

Project management for Large and Complex Projects is not like manufacturing. It is not predictable. It is not about Six Sigma or maintaining variances as low as possible compared to the plan. It is about acknowledging that change is part and parcel of Project management, and designing processes and systems around it. It is about identifying what is the real purpose of the Project, and subordinating the plan to reach this purpose. Change is not an exception in Project management. It is part of normal life. Let us design Project delivery processes that are agile and built around change management.

Stop moaning about changes to the plan! All experienced Project practitioners know that it is the thrill of finding solutions to the most unexpected and intricate situations that make the interest of the profession - and why they have chosen it. Change is part and parcel of the fun of Project leadership. Let us recognize it as such, as the daily challenge of any Project practitioner, and build our processes around it.

Chapter 11: Increasing Schedule Reliability

Chapter Key Points:

- Project schedules are always optimistic (as a rule of thumb, by 15% to 20%). Make sure you have some buffer!
- Psychological effects play a very important role such as Parkinson's rule, the Student syndrome, the commitment syndrome, the "planning fallacy" and the general lack of calibration of estimates.
- Introducing buffers in the schedule is essential and needs to be done properly to avoid project execution issues.

How To Be On Schedule Reliably

There is a practical joke in the airline industry. It could have been applied to maritime navigation too. The airlines that are statistically the most on time as shown by consumer reports... are often those that add a contingency buffer in their schedules. It is easy to see: compared with competitors that have the same equipment, their flight schedules for the same destinations are a few minutes longer!

That means that if on time reliability is important for you, and it should necessarily be for a Project Manager, a good way to achieve it, on top of operational excellence, is to add a little bit of reserve for the unexpected in the initial plan.

Unfortunately, Project schedules are always somewhat optimistic – for different reasons that we will debunk first. Then we will examine how we can use contingency in a useful way to make sure the Project will be more reliably delivered on the expected time.

As we'll see in this Chapter, a significant part of the art of developing schedules is to include buffers at the right places to deliver on time while keeping the activity durations challenging for the contributors.

Why Project Planning Is (Almost) Always Optimistic

Conventional Project schedules are fundamentally optimistic because Project planning almost never accounts for:

- inefficiencies in the handover of tasks between contributors,
- intrinsic complex issues like workplace congestion and similar constraints to simultaneous works (e.g. when working on different levels of the same facility),
- effective coordination of contributors,
- resource multitasking (between tasks on a single Project or between different Projects), etc.

At the same time we know from ample literature that tasks durations are often exaggerated by those responsible for them as they feel it is a commitment on which they might be judged later. This psychological effect to 'pad' one's estimates of the duration needed to do the job might seem to be a factor that could compensate the relative optimism of Project planning. Yet we also know that once a task duration is entered in a schedule, activities tend to fit within that timeframe (through such effects as the Student's effect – people start at the latest possible moment, Parkinson's law – work tends to fit the time available - or simply because people lack of incentive to be effective when they feel that they have time to do a particular activity). Finally, statistically, these effects imply that tasks almost always take more time than what was entered in the baseline schedule. The optimistic nature of conventional Project planning still holds in practice.

Major psychological factors at work in project schedules

- **Parkinson's law:** work tends to fill the time available [if the task if planned for longer than it would take, work will still take at least that duration]
- **Student's syndrome:** if people have time to do a task they will always wait for the last possible moment to start
- **Commitment syndrome:** people will always 'pad' their duration estimates when they are asked (consciously or unconsciously) to commit to a duration. Hence they announce durations that can be much longer than what is achievable.
- **Planning Fallacy:** an optimistic bias on the duration of own's future tasks, irrespective of benchmark data on past duration distribution of similar tasks (Daniel Kahneman, Amos Tversky)
- **Lack of calibration of estimates:** when people have not calibrated their estimates comparing to actual durations, they will tend to be pessimistic (conservative) for usual tasks and optimistic for unknown tasks.

Rule of Thumb for Delays to be Expected in Complex Projects

At Project Value Delivery we use the following rule of thumb for delays in complex Projects: we consider that **the basic delay that can be expected is generally on the order of 15 to 20% of the initial Project duration** for a typical Project plan.

The following factors will diminish this estimate:

- Low complexity of Project schedule (in particular, low value and complexity of procurement, and number of Project offices),
- Existing possibility/margin to compensate delays through specific effort or additional resources (Project office, construction site),
- Existing lessons learnt on similar Projects with actual durations,
- Existence of buffers on critical chain (diminish basic delay estimate by the duration of the buffer).

The following factors will tend to make this estimate be a minimum ballpark:

- High complexity of Project schedule, in particular regarding complicated procured items and complex logistical arrangement; and regarding split Project offices with poor communication,
- Poor resourcing of key manpower-intensive activities that do not allow for any visibility on the available margins for additional effort should key activities fall behind,
- No convergence monitoring (including reliable schedule update).

This rule of thumb of 15-20% basic delays might look hugely significant but it corresponds often to the reality of Large, Complex Projects – which can get much worse once they start getting astray. At some stage in the 1970's in the North Sea, at a time when the associated technology was not entirely mature yet, offshore Projects were routinely costing and lasting 3 times the initial estimate. In complex Projects involving new environments or cutting edge technologies, a ratio of 7 to 10 is often observed[2].

The consequence on cost will ensue from the schedule delays, often with somewhat less acuity because procured items keep the same value (although procuring them urgently due to delays in engineering will often incur some additional costs). Major sources of additional costs are to be expected from the manpower (including contractors) and equipment, including Project office and construction site work. Cost overruns to be expected are thus, as a rule of thumb, on the order of 10-15%. This is higher than the contingency taken by Contractors – often around 5-7% of the cost at contract award- but reflect quite well the actual usual final outcome of most Projects once all Change Orders and other claims have been settled. After all, changes from the Owner/Client and changes due to interface issues between parties are also part and parcel of Project complexity!

---

[2] Those mathematicians amongst the readers can use the well-known heuristic of a multiplier of π for projects that don't go well and π^2 for projects that really fail.

List of Factors that Influence the Impact of Conventional Planning Optimism

In summary, here are a list of factors that are favourable for recovering from the intrinsic planning optimism, and those factors that make the situation worse:

Factors favourable for recovery	Unfavourable factors for recovery
Limited complexity – linear Project	High complexity, highly converging Project with many different contributors (in particular complex supply chain of complicated equipment)
Convergence monitoring and usage of buffers in front of main convergence points	Poor convergence monitoring, no buffers in schedule
Schedule is realistic for manpower-intensive phases of work through good resourcing and there are resource margins for acceleration / compensation of poor productivity (e.g. possibility to go in 2x8 or overtime)	Poor resourcing of schedule for manpower intensive phases of work. No realistic assessment of required productivity or mobilization rate for resources.
Schedule is focused on the key activities, has the right balance that allows good quality update and change agility	Highly complicated schedule with high detail that requires lots of resources for update or change
Key activities duration ascertained from past Projects' similar activities, not from opinions	No estimate backup as to the duration of key activities
Possibility to count on Project team to work overtime to fix critical Project office deliverables, or ready availability of additional competent resources	Rigidity of Project team productivity, resource level, and organization

Why You Need To Buffer Your Schedule

What is the Impact of Project Planning Optimism?

Because conventional Project planning is fundamentally optimistic, reality is always worse. Activities are late. The 'virtual' float syndrome further gives time and creates delays throughout the Project. To mitigate the situation, tasks get reassigned, sequences of tasks reworked.

Of course those tasks which are particularly important are those on the Critical Chain (the critical activities and resources that drive the Project delivery). For more than 90% of the Project tasks, delays on individual activities have a weak impact on the overall Project outcome because they have intrinsic float by not being critical. However it is important to identify those activities which are normally not critical but might become so, because they might suddenly change the focus of the Project.

Why Couldn't the Approach to Protect from Cost Variability Be Used for Schedule?

In terms of cost, probably due to the pressure of competitiveness, practices have been adopted across the Project industry to deal with cost variability. Cost line items are not padded individually, and to protect the Project budget, a single contingency element is added to the budget. This contingency element can be derived using several methods, the most advanced of which is using Monte Carlo simulations (refer to our *Cost Control* and *Project Risk Handbooks*).

The concept of contingency in place of individual cost padding allows the total budget to remain competitive while providing a sufficient amount of protection against cost variability during Project execution. In that logic, cost items have to represent achievable (P50) values. P50 means that there is a 50% chance that the cost will be lower (and thus 50% chance it will be higher). Conversely, a P80 means that there is an 80% chance that the cost will be lower (and thus 20% chance it will be higher). A padded estimate will typically be a P80, P90 or P95.

Contingencies in large, complex Projects are generally on the order of 5-7% of the total budget, provided the estimate is recognized to be of a high quality.

An important element of the contingency approach is that the contingency belongs to senior management (above the Project Manager) and cannot be used without their permission. This ensures that cost discipline is indeed applied throughout Project execution.

While the contingency approach is very mature in cost management, it is quite amazing that the same approach is not used so much for schedule. And this happens although for Large, Complex Projects, schedule is obviously the main driver of the Project performance!

Conventional Project management practice thus still considers a schedule which is a linkage of tasks, which are given a certain duration, without any apparent explicit treatment of their potential variability. It is even widely recognized that task duration variability is treated implicitly by padding the duration estimates. There is no time contingency added to the schedule. Why is this practice commonly accepted in the field of schedule management, when it is recognized to be a poor practice in the field of cost management?

Moreover, we now know that in the field of schedule, this practice leads to a psychological vicious circle: confronted to more time to do a task than absolutely required, the responsible person will generally fall into Parkinson's or Student's syndrome and will miss the target after having shown a very low productivity at the beginning of the task.

It is thus important to recognize that padding time estimates is a poor practice and that time variability needs to be treated explicitly and separately, exactly as it should be in cost management.

Introducing Buffers as Contingency

Good practice in scheduling thus would involve buffers acting as schedule contingency at the back of the activity sequences, and in front of the main convergence points and in front of the final Project delivery. The duration of these buffers need to be owned by the Project Manager or above. Just as cost contingency, their evolution with the progress of the Project can be monitored.

For the sake of clarify we make a difference between buffer and float. Buffers need to be identified as activities in the Project schedule, which duration is managed by the Project Manager. Float is a quantity that can be computed but is not represented by an activity in the schedule. A buffer is thus, in a way, a float that would have been materialized in the schedule.

Buffer	In this book, a duration that is under management control to constraint parts of the Project schedule and/or to protect completion dates. Buffers are represented by mock activities in the Project schedule, or hidden in the duration of existing activities and they are directly under the control of Project management.
Float	Amount of time an activity in the Project schedule can be delayed without impacting either the subsequent activities (free float) or the Project completion date (total float). In this book, Float generally refers to total float for the sequence of activity under consideration. Float is a value that is calculated, but is not represented physically in the Project schedule.

The following figure summarizes this approach in a graphical manner.

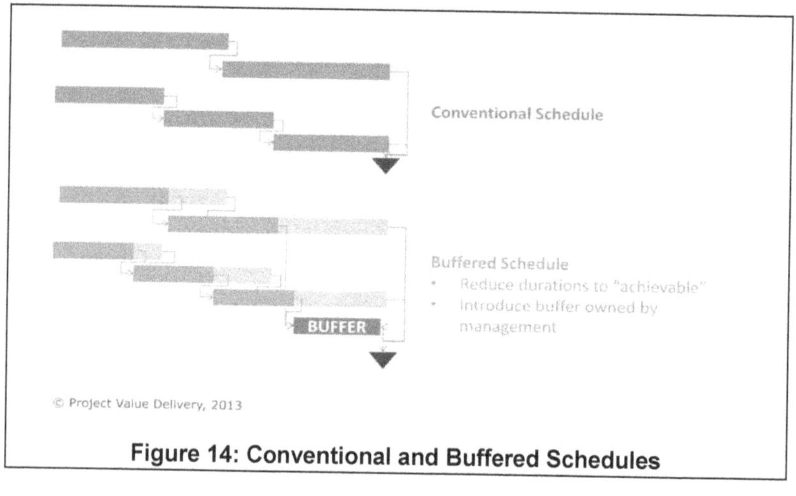

Figure 14: Conventional and Buffered Schedules

As we can see in the figure, the removal of padding on activities has also the advantage to identify more clearly what are the actual critical activities that will drive the schedule (in this case, the second sequence). This was hidden or even inaccurate when the activities' durations were padded.

As mentioned in Chapter 9, a more advanced buffer usage is to introduce local buffers to avoid the effect of 'virtual float' creation and maintain the tension on the Project. When activities get delayed tension may be released on other activity sequences. The best practice is to avoid this by including a new buffer that compensates for the added float, thus maintaining the schedule, and putting the usage of any additional time under the permission of the Project Manager.

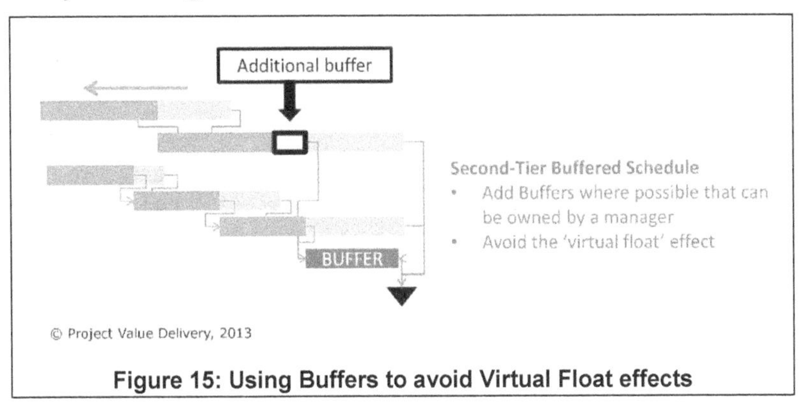

Figure 15: Using Buffers to avoid Virtual Float effects

How to Buffer Your Schedule in Practice

Setting explicit buffers in schedules is not a widely accepted practice because it creates contractual issues in a Contracting environment. In particular, the ownership of the buffers may be disputed if they appear explicitly on a schedule between Owner/ Client and Contractor (see discussion on the ownership of float in Chapter 12). Hence in most cases the Contractor needs to hide the buffers inside some activities – it is common to have them part of back-end activities such as commissioning. It is also possible to make them apparent in non-challengeable locations, such as buffers:

- Placed before mobilization or other key convergence points,
- To account for weather contingency and other external constraints,
- To account for certification and regulatory issues.

What is important is to make sure buffers are included, to be aware of the duration of these buffers even if they are not shown and manage them properly during project execution.

Introducing Buffers at Schedule Development Stage

An approach similar to the cost management method should be used in scheduling. Activity durations should be set at the P50, without padding. Any "allowance" for the duration of a particular activity should be identified explicitly and assigned to this particular activity. The Project delivery date should be protected by a Project buffer, which is an overall "contingency" applied for schedule.

As in the best practice in cost management, at the beginning of the Project, no activity buffer should normally be allowed, only an overall Project buffer. This is to ensure that the baseline Project schedule takes into account the P50 durations.

How to Use Explicit Buffers During Project Execution

The most important in the usage of Project buffers is that they should be 'owned' by the Project Manager and Project team personnel should have to ask permission to use part of it to offset some unexpected events. This gives back control to the Project Manager and allows him to use these buffers in line with the Project purpose. It is a significant advance compared to the traditional situation where there are implied buffers in the schedule created and used by contributors without knowledge or influence from the Project Manager.

Project buffers allow a healthy monitoring of the effective convergence of the Project. During the Project execution, the evolution of the Project buffers can be monitored and corrective measures taken early if this evolution becomes divergent, way before the delivery date of the Project is impacted. Project or gate buffer monitoring can be done similarly as described in Chapter 9 for float monitoring.

The main benefit of the buffer method is to have at Project Manager level an accurate knowledge at all times of the conservative assumptions that have been introduced in the Project schedule. These can be shared and discussed explicitly with interested stakeholders.

A secondary benefit is a much better usage of opportunities. In conventional scheduling practice, it is very rare that opportunities linked to shorter than expected durations of activities, if they arise, are effectively exploited for the Project. By making sure that the time estimates are not padded, the baseline schedule is more aggressive which allows for a better utilization of the possible opportunities – while protecting the delivery date with a buffer. Leveraging opportunities when they arise can be further enhanced by identifying the critical resources involved in the critical activities and making sure they are ready to work even if the previous activity finishes earlier than expected.

How Can We Estimate the P50 for Activities?

One of the greatest challenges is to estimate P50 durations for activities, because Project team members and suppliers alike will have a tendency to pad their estimate to ensure that they fulfil their delivery commitments.

A strong database is needed to be able to benchmark actual durations. Such a database often exists for cost and not for schedule; it needs to be extended to schedule data as schedule is indeed the main driver for Large, Complex Projects.

On the short term, while this knowledge database is being developed and filled, some rules of thumb can be used. For documents, engineering deliverables that are done in-house, and other cognitive work, P50 can often be taken as short as 50% of the announced duration. It seems harsh but is a very valid rule of thumb. When you introduce the method in an organization you might want to start with a less harsh rule and reduce announced durations by 25%-30% only.

For suppliers, duration estimates can be made part of the competition in parallel to price so as to have a feel of what is reasonable to expect compared to what would be announced without such constraint. Actually in case where there is little downside for being late at delivery, suppliers might tend to provide excessively aggressive schedules.

How Can We Estimate the Project buffer?

A rule of thumb would be to have a Project buffer equal to 10% of the Project duration found by using P50 estimates for a simple Project with low convergence pattern, up to 25% for a complex Project with a high convergence pattern.

Conclusion

In large and complex Projects, schedule performance of a Project is directly linked to its overall performance. A schedule should strive for 'achievable' activity durations and include a single contingency buffer under control of the Project Manager. It is vital to make the assumptions about contingency buffers as explicit in the schedule as they are in the cost. This Best Practice, combined with the Convergence Plan monitoring and Float Monitoring techniques, has multiple advantages over the conventional scheduling approach, while building on it in an easy manner. It increases the robustness of the schedule dramatically, keeps the project contributors under pressure, while giving the Project Manager additional control tools that he can leverage to keep control of his Project.

In the end, for complex Projects it is absolutely necessary not to be too ambitious and optimistic in the schedule. This might be difficult in contractual contexts where a short schedule could be an argument for competition. Should it happen – which is the case more often than what would be recommended, even vis-a-vis internal senior management, Contractor and Owner/ Client alike must expect necessarily the Project realization to be delayed compared to the optimistic schedule developed by and for salespeople. As part of that game, Contractors need to ensure there are appropriate opportunities in the contract and the Owner should ensure it has a sufficient reserve to cater for the inevitable growth both in terms of costs and delays. This too usual approach however is not recommended and a sounder approach involving schedule buffers would be a more mature way to address the issue of Project on-time delivery. Competition could still be organized on the basis of a schedule without buffer, and a contractual standard buffer added on top at contract award, that could be for example 10% of the estimated Project duration, varying based on the Project complexity.

Chapter 12: Contractual Aspects of Schedule Management

Chapter Key Points:

- Time is of the essence in Project execution.
- Time has got as much value, if not more than cost.
- The Contract should be reviewed thoroughly to understand what are the obligations and possibilities.
- The float should be owned by the Project as much as possible. This should not prevent the Owner/Client of having its own time reserve beyond the Project completion date.
- Schedule updates need to be rigorous and documented, and warnings of delays need to be communicated formally between Parties to be traceable.
- Extensions of Time are always difficult to quantify and justifications are often controversial.
- The Simplified Project Schedule is an excellent tool to support management-level discussions regarding Extensions of Time and changes of execution strategies.

Time is one of the three composing factors at the root of any large complex Contract, together with cost and scope. As such, it is often said (and sometimes even stated in a contract clause) that "*time is of the essence*". This encompasses the fact that schedule compliance is at the source of the agreement set forth in the Contract, including for a start, a fixed target for completion, and mechanisms to enforce the responsibilities of each party in regards to its achievement. It also highlights that decisions must be timely to minimize the impact of events.

The subsequent sections are written in general terms; in all cases the particular Contract provisions under which a Project is executed need to be studied carefully to understand the specific requirements applicable to a particular Project.

Never Forget that Time has More Value than Cost

It is important to remember that in all Projects, time is extremely valuable for the Contractor (due to the standby costs implied throughout the entire Project, and other indirect disruption costs, plus the potential for Liquidated Damages), and even more for the Client (due to delays and loss of production which can impact substantially the economics of the Project, as well as create possible claims from other parties). Hence time has a very huge intrinsic value because of indirect impacts. Claims related to the consequential effects of time will often have more value than claims for direct additional costs.

> **Time has always more value than direct costs.**
>
> **Never forget to claim for time!**
>
> — (Project Value Delivery)

This has the following consequences that all Project Managers need to understand fully:

- From the contractor perspective, any change by the Client of a time-bound milestone, Owner-provided deliverable, or mobilization window, has a very significant value that has the potential to be claimed either as an Extension of Time (EoT) or as an additional cost (which needs to take into account direct and indirect costs, such as Cost of Opportunity),

- Delays, standby (or suspension) has an intrinsic direct cost and substantial indirect costs thus a very significant value related to the time lost, that will be claimed systematically by the contractor and need to be reasonably bounded from the Client perspective,
- Clients need to be wary of delays created by poor performance of the Contractor such as poor productivity, inadequate overall scheduling and poor anticipation of long lead activities; this explains the emphasis put in verifying that the Project schedule is realistic and well-built including the quality of reforecast taking into account observed productivity,
- Project Managers need to be wary of situations where time delays snowball, which makes it even more important to note and evaluate the first delays as soon as they occur to achieve a full traceability of the cascading events.

Contract Review: Identify the Particulars of Your Contract

Even if the principles ruling the concept of time are largely shared across all Large Complex Projects, a complete understanding of the specifics of the Contract needs first to be established by the Project team. This is normally part of Project start-up through the exercise of a comprehensive Contract Review.

In addition to locating all time-related clauses in the Contract (which shall prove useful in further exchanges with the other Party), it will also allow the Project Management Team to gear the establishment of procedures and baseline schedules. This then allows communication of such targets to the Project team.

The particular points of interests of any given Contract – although maybe spread out in various clauses – shall be the subject of a checklist during Contract analysis:

- The start date of the Project, with in particular clarification between Contract signature date, effective date, and the start date of the countdown towards project completion, which may be one of those two or also commonly the date of site Handover, of financial closure (Final Investment Decision), of project validation by stakeholders (Initial Project Validation) or even an earlier date than those in integrated projects, like start of feasibility studies, Open Book Tendering or of a Memorandum of Understanding pre-empting the complete Contract,
- The final Completion Date(s), which is linked with scope clarification: Ready For Commissioning, Ready For Startup, Initial Acceptance (after performance tests), submission of an Estimate, etc. This date can also be split for Projects calling for completion in parts or in phases,
- Liquidated Damages clauses, pre-defined amounts of money designed to compensate the Owner for the annoyance and losses due to the delays to one or several of the Completion Dates, without having to demonstrate such loss. This compensation might not be commensurate with the losses from the Owner but are generally designed to substantially impact the Contractor's profit and thus create a strong motivation on the Project delivery date,
- Overall schedule, generally in the form of a list of main key milestones or simplified schedule, which are not mandatory as such or linked to Liquidated Damages, but are often the framework for clauses binding the Contractor to launch recovery actions, or triggering acceleration, penalties, incentives, or options exercise,
- Remaining obligations after the Completion Dates, therefore out of the scope of the main Contract and beyond the Liquidated Damages' reach, but still important to be kept in mind: spare parts delivery, final as-built documentation, warranty period, performance tests, site reinstatement, etc.,
- The external interfaces inputs, generally in the form of elements due at a definite point in time by the Owner or its affiliates, failure of which would give

entitlement to the contractor to claim for extensions: site access, design dossier handover, free-issued materials or equipment, availability of utilities or infrastructures, etc.,

- The mechanisms and lists of events allowing for Extension of Time (see sections below),
- Any exclusion or specific constraints applicable.

One general rule of law which needs to be highlighted is that in the absence of such clauses or details, the only binding element will therefore be the law of the Contract. For example in the common law jurisprudence:

- if there is no completion date or no mechanisms in the Contract to decide a new Completion Date if the initial one is obsolete, then it is said that 'time is at large', and the only obligation of the contractor will be to complete within a 'reasonable' time,
- without a Liquidated Damages clause, which has the advantage to clearly extinguish the debt due for delay, the responsibility of the contractor is not limited, and the Owner may very well claim for consequential losses disproportionate to the contractor's Project value.

Therefore it is essential to go in detail through a complete Contract Review and be wary of the constraints applied to the Project schedule, and also of the underlying missing clauses and associated risks.

Who Owns the Float?

In contractual discussions related to schedule, the respective responsibility of the Owner and of the contractor is always a matter for debate. These issues have created a large jurisprudence which evolves constantly, and is the subject of many contract management books. We will here discuss only some fundamental principles commonly accepted. Again, these need to be reviewed with the particular Contract's provisions in mind, which might contradict those default general principles.

One such subjects is the ownership of the float, i.e. the time lapses between activities or at the end of the project, which allow some buffer, and which can be used to cushion delays without impacting completion dates.

We believe that for lump sum contracts or equivalent, a general rule should be that the Total Float should belong to the contractor up to the point of showing it plainly in the Project schedule, in order to protect it. As a matter of fact, the contractor takes the responsibility of a completion date in exchange for a share of the value of the contract, which is part of the risk apportionment in the agreement.

The float shall therefore be built in schedules actually as a contingency buffer to protect the Project's Completion Date (ref. Chapter 11). It is important to include this issue in the contractor's contractual principles, and in the establishment of the schedule, by explicitly including for time allowances at certain critical or highly uncertain steps of the Project. Instead of hiding reserves, we trust that it is a good practice to show them clearly as part of the contractor's entitlement and execution plan, and then manage openly their evolution and consumption.

This methodology does not deprive any party of the opportunity to use the existing float, but allows to do it while being conscious of the additional risk burden imposed on the Project, of the costs linked to it, and possibly of the loss of opportunity. It will ultimately render demonstration of Extensions of Time much clearer and easier for all parties. An example is shown in the following figure, where a delay attributable to the Client is followed by a delay from the Contractor. Without visible buffer, the Contractor takes the full responsibility of the delay; with visible buffer, it is much clearer that most of the delay is in fact due to the Client.

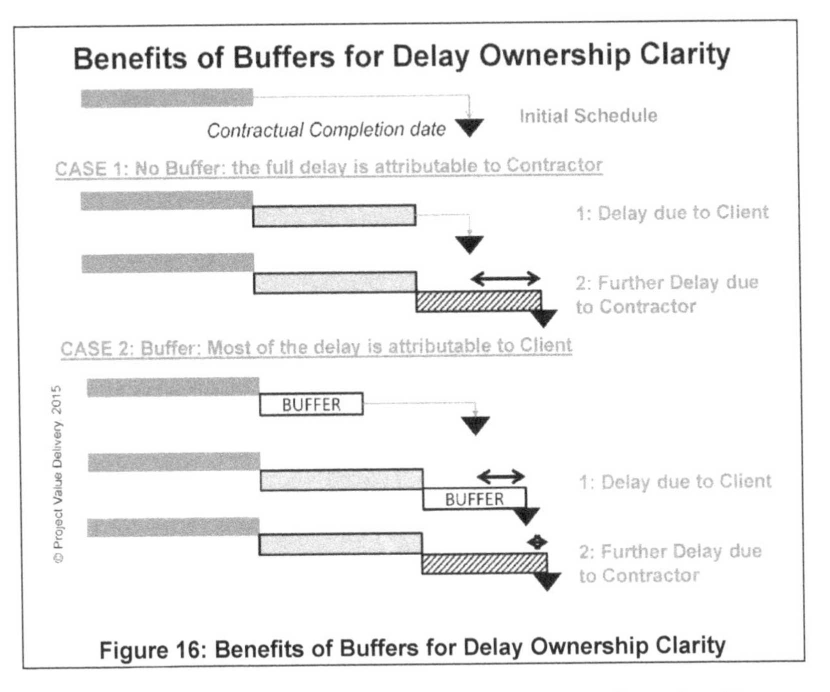

Figure 16: Benefits of Buffers for Delay Ownership Clarity

Conversely, the Owner should organize itself to maintain a separate "Owner reserve" beyond the contractual completion date if that is critical to its own operations (such as an interface with another contractor), just as it generally happens in the area of cost.

Justifying and Demonstrating Extensions of Time

The central mechanisms of any Contract are designed to respond to the need of agile Project Management: all Projects are prone to change. In some cases, these changes are constructive like the release of an option, a change in construction method or equipment, etc. In most of the cases however change is the result of unexpected or unaccounted for events, and the need for a hasty patch-up solution. In terms of schedule, Extension of Time (EoT) is the contractual process allowing completion dates to be altered within a pre-defined framework.

Many methods exists as to how to present and demonstrate an Extension of Time, but a central logic always prevails: faced with an event impacting their capability of meeting Completion Dates, Extensions of Time are meant to ensure the return to a contract equilibrium, i.e. the fair balance of responsibilities between the two parties, roughly equivalent to the balance agreed at its signature.

Main Reasons for Direct Claims of Extension of Time

While the accepted reasons for possible claims of Extension of Time vary from Contract to Contract, all changes related to the Client or other stakeholders that would not be under the control of the contractor would be candidates for a potential claim.

This typically includes:

- late drawings and technical specifications from the Client (e.g. fluid specifications, soils surveys etc.) – this also includes later updates and revisions if they bring changes,
- late free-issued items delivery from the Client,
- late regulatory authorization delivery when these authorizations are under the responsibility of the Client,
- unforeseeable adverse physical obstructions or conditions, including worksite interface problems with other Contractors,
- suspension of the works ordered by the Client or the authorities,
- failure to give access to the worksite,
- additional or extra work,
- exceptionally adverse climatic conditions (claims under this aspect are generally very strictly limited in contracts, such as only for named storms and cyclones),
- any delay, impediment or prevention by the Owner within specific bounds (for example, suspension of the works due to faulty HSE conditions or any situation where the Contractor will not have showed the basic expected workmanship will not be a valid cause for a claim).

The rest of the Extension of Time clause is usually establishing a framework to define the rights and responsibilities of each party as regard their contractual treatment: requirement for notices and supporting analysis and documentation, time limits, recourse in case of dispute, etc.

However good the provisions of a Contract are, the discussion about and award of an Extension of Time are always been known to be very controversial - much more than cost. The Project Manager needs to proceed with extra preparation and caution on this matter.

Intrinsic claims for Extension of Time

When the Contract includes both reimbursable/ time-rate elements and lump sum elements (such as for example, when the contractor did not want to take the risk of certain difficult operations, or simply for weather standby issues), it is important to understand that the extension of duration for the reimbursable operations beyond the initially expected duration should normally be systematically reflected as an extension of the due Completion Date of the Project. Hence even without any ground to a claim per se, the Project Manager should file a claim for Extension of Time in case of delays to a time-rate portion of the work to make sure that the remaining works will not be squeezed with the danger of being hit with Liquidated Damages. This reminder is important as it has led to many disputes.

In case of extension of reimbursable/ time-rate works due to external reasons, never forget to claim for an Extension of Time to the final delivery date of the Project.

Methods to Demonstrate an Extension of Time

Contractually, it is important to be reminded that the claimant bears the burden of proof: in no case will allegations or a non-supported case bring any recovery – it is therefore mandatory to build a strong intelligible argument and support of this proof. It follows that the main methodologies used to demonstrate Extensions of Time effectively are the same ones needed to defend against them, and go through three main components:

- Proper segregation,
- Accurate demonstration,

- Contemporaneous documentation.

Proper segregation

The primary element of a good demonstration is clarity through the segregation of the elements building the case; hence answering to the question 'what is the story?'

This story, to stand together, shall be built around five elements, which shall be a reflection of the contract Extension of Time clause:

1) The **Delay event shall be identified**, and singled out,
2) The **Delay liability shall be proven** through reference to Contract clauses, ascertaining the delay event is compensable,
3) The **Delay causation shall be established**, to clearly show that the impact directly and undoubtedly follows from the delay event,
4) The **Delay impact shall be ascertained**, qualifying what consequences are at stake deriving from the initial delay event, direct or snowballing effects,
5) The **Delay compensation can be claimed**, quantifying the reparation sought through the provisions and rights set in the Contract.

The segregation of those elements is a first task which will help shed light upon the issue at stake and help the production of a clear and defendable reclamation.

Accurate demonstration

The technical and central part of an Extension of Time request will be the forensic schedule demonstration, which does the job of linking the delay event to its impacts. As per the requirement set above this demonstration will need to be extremely accurate and defendable. Largely accepted and highly regarded are the SCL Delay Protocol and the AACEI Recommended Practice 29-R03, in which the main technical ways for analysis are compared and detailed.

Two main forensic demonstrations can be sustained, according to the conditions and the interest of the Parties:

- a predictive analysis forecasting a future impact
- a retrospective analysis, reporting a state of fact.

The retrospective analysis has the advantage of showing a fixed final impact status already taking advantage of possible mitigations and recoveries, and by nature will be recapitulative of all actual impacts, allowing for a very strict quantification. It is therefore often preferred by Owners. While the quantification of the impact will be better as it occurs in hindsight, on the other hand this approach makes more difficult the attribution of responsibility of the delay, which will require forensic methods and extensive analysis and debate.

On the contrary, a predictive analysis is the tool of choice of contractors. It allows:

- an easier attribution of responsibility,
- claiming future impacts on the account of contemporaneous events,
- seeking to re-establish the contract balance immediately,
- clarifying the level of risks carried over, and
- not taking the risk to wait for an overall settlement at the Contract completion.

The drawback is that the consequences will be assessed on a forecast that might not reflect what will actually happen, thus leading to some subjectivity in the impact assessment discussion.

Five main tools can be used for the purpose of an accurate demonstration, sometimes concurrently:

- Predictive tools:
    - **Time-Impact**, an analysis done at the time of the occurrence of the event, simulating future results of the event(s) on the uncompleted portion of the project and Completion dates, on whole or part of the schedule network,
    - **Impacted As-Planned**, a simulation done on the initial baseline schedule, with insertion of the delay events, to model out what would have been the result of said delay, all other thing being equal,
- Retrospective tools:
    - **As-Planned Vs. As-Built Comparison**, a simple comparison between the initial baseline

schedule, and the final as-built schedule, after Completion has been reached,

- **As-Built But-For**, the reverse simulation, done on the final as-built schedule, to which the impacting event is subtracted,
- **Progress and statistical analysis**, showing progress shortage, and relative prolonged activities or trends on completion, attributable to the delay event(s).

Finally, in terms of scheduling techniques, every planner shall keep in mind that Time slice/ Sub-networks/ Fragnets / Window analysis can be used, to capture more clearly the event and its impacts along the Critical Path(s) and to avoid indulging in a too complicated demonstration. This generally allows a clear segregation of the delay origins.

Do not pace your progress with regard to events.

While establishing a demonstration, it is important that work progresses steadily: although it seems like a cost-effective action to spread out the works, this is also a risk of engaging oneself in culpable delays if the extension if finally not granted or is related only to a specific part of the works.

A contractor is said to be "Pacing" his progress when he slows down activities in anticipation of an Owner's delay. This is a contractually dangerous action which will be very difficult to justify if it has not been announced to the Owner prior to its implementation, and received Owner's agreement or instruction. In general, it is not recommended to lower the pace of Project activities as it is always better to be early than late, and delays from the Owner can release resources to deal with other pressing issues. It will also make the effect of the Owner's delay very clear when it will occur, thus supporting a further claim for Extension of Time.

Contemporaneous documentation

Lastly, it is of extreme importance to tackle the subjects contemporaneously, and with the proper amount and accuracy of supporting documentation. This will anchor in reality and through facts the dates and constraints justifying the impacted schedule.

The documentation will be the basis of the argument and the support of the proof. It can take many forms:

- pictures of the obstructions on site,
- delivery reports showing the actual date compared to a contractual one,
- site daily progress reports stating events and signed by both Contractor and Owner representative,
- letters of instruction,
- quality control tests requiring reworks,
- quotation of a vendor for additional works or materials,
- etc.

This documentation needs to be gathered at the time of identification of delay events, as it tends to be much more difficult to gather it afterwards, even after a few weeks, and even more so at the end of the Project.

This documentation is best captured early and formally. The sending of notifications and trends analysis, often required by Contract, will also serve the common purpose of cooperation between the parties in identifying and hopefully proactively mitigating risk or potential delays area. A proper communication protocol will give an option for recovery actions to be undertaken, and if unsuccessful, will provide a solid Extension of Time file to demonstrate the causes, entitlement, goodwill and remaining impacts if the actions have not allowed full recovery.

The gathering of support documentation must be accurate and allow backing up the entitlement demonstration in a stand-alone manner. It has to be sufficiently accurate and decisive (not assuming or inferring, but asserting) to ultimately be able to convince a judge or arbitrator, who will have no knowledge of your case, except the Contract and this specific documentation.

Missing or improper documentation will make it equally difficult to convince the other party beforehand.

Disruption claims

Beyond single events impacting the schedule of the Project, the issue of disruption by the Client creating contractor productivity issues is often an area of deep controversy. It can be extremely difficult to have evidence of such disruption, in particular because of the pervasive impact of disruption that cannot be easily segregated. The only methods available to assert a valid cause include:

- Comparison of turn-around time of documents with the contract requirements,
- Comparison with well-established productivity benchmarks in similar conditions,
- Evidence of Client representative's intervention beyond its reasonable scope, or of transmittal of additional comments beyond the comment deadline etc.
- Comparison of the productivity between two time periods during the Project, when disruption is only claimed for a particular period.

Communication and Demonstration

The preceding sections covered the actual detailed demonstration of Extension of Time claims. This generally leads to very detailed discussion between scheduling and Project professionals. At the end however, the decision will have to be taken by Senior Management, generally at the sponsor level or even higher for substantial claims.

Notwithstanding the need to build a complete file that can sustain expert scrutiny, the case needs to be distilled in a few considerations that are simple to understand. We strongly recommend the use of the Simplified Project Schedule as a key vector to demonstrate to Senior Management what has happened in the Project, the fundamental cause and its consequence (the Convergence Plan can also be used to highlight substantial changes to the plan but because activities are not linked, it will not have the same impact). In addition the Simplified Project Schedule can easily be subject to what-if analysis and even to Schedule Statistical Analysis if needed to assess the resilience of some options.

At some stage the quality of the communication is essential to lead to conviction. At Senior Management level, be sure to use the right vector for your arguments. As a support, the Simplified Project Schedule is the ideal candidate.

Cost Valuation of Time

The quantification of additional costs due to delays can be very difficult, as different impacts and perspectives need to be untangled to determine a right and fair compensation.

A Contractor and an Owner have initially agreed upon a Completion date for a certain price. A consistent valuation needs to be done of the delays when due to impacts of the Owner. These delays:

- cause direct costs,
- deprive the contractor of their allowances for internal delays,
- remove the contractor's liberty to take mitigation actions freely by varying the workforce and construction methods.

The latter effects can be calculated as the price of the uncertainty and can be quite visible at a bid stage, when a selected Contractor, asked to reduce his promised overall duration of works, increases the contract price.

Categories of Delays

In terms of cost quantification, delays subdivide in categories:

- **Excusable Delays** are delays that are unforeseeable and beyond the control of the contractor, according to the Contract particulars,
- **Non-Excusable Delays** are delays that are foreseeable or within the contractor's control.

Obviously, the distinction between these two is significant in that it determines which party is liable for the delay. If it impacts the Critical Path of the Project, an

Excusable delay creates contractually a valid excuse not to achieve the Completion Date timely. It dictates entitlement leading to an Extension of Time (financially speaking the relief or postponement of Liquidated Damages).

Excusable Delays are further subdivided into two categories:

- **Compensable delays** crystallize the events in which the normal march of the Project is impacted (on the Critical Path or not) by causes in the control of one of the parties, creating an unbalance compared to the initial Contract terms, for which the injured party may claim recovery to the other,
- **Non-Compensable delays** will be the cases in which neither the contractor nor the Owner has control (or accepted risk responsibility in the Contract) over the cause of delays (for example weather impact, strikes, etc.), and therefore the fairest compensation is that both parties assume their own additional costs.

Concurrent Delays

Beyond the question of whether the event stated by the contractor for an Extension of Time claim is receivable under the Contract, the main issue will always be whether the delay is effectively completely due to that event or whether this claim aims to cover some delays that the contractor has suffered due to poor performance of the activities under its control.

The general rule is that the Extension of Time to be granted should only be the actual additional delay due to the event at the origin of the claim, beyond the delay that is due to the inherent contractor's performance.

Hence, in case of concurrent delay, only the additional delay to the Project Completion Date due to the Owner or to an acceptable external event is receivable. The situation obviously complicates if there are some indirect effects from the Owner's delays on some other aspects of the work which could cause additional Contractor delays.

It is generally up to the Contractor to identify and evaluate concurrent delays. The Owner needs also to be in a position to challenge the Contractor's assessment. In terms of costs, Extensions of Time will always give rise to a relief of Liquidated Damages (Excusable delay), even if a concurrent delay occurs, as the rule of law is that the Owner shall not take advantage of concurrence. Additional compensation will have to be apportioned and calculated on the culpable events impact only.

Direct and Indirect Cost Impacts

To be in a position to properly quantify the value of time for compensable delays, it is important to categorize the different unbalances that can arise out of delays, and how costs will follow the type of impact to be rectified.

- Direct costs of delay will be quantified quite simply based on contract rates or on invoices for standby (manpower and equipment made idle for a period of time), additional tasks (like reworks on portions of work already terminated), extended tasks (work being spread out, creating supervision and running costs increase for the same productive output) and disruption (loss of productivity). Care should be taken to include the associated additional running costs for the Project Management Team and associated overhead,
- Indirect costs are the costs not directly billable but which affect the project's margin or the risk profile:
    - relief or application of Liquidated Damages,
    - Cost of Opportunity in possibly losing other Contracts when some unique assets (crane barge, high capacity crane, some specialized staff, etc.) are to be used beyond plan,
    - allowances reduction through the reduction of float and the possibility to mitigate some delays at low direct cost
    - possibly reputational costs, for particular Projects where not reaching Completion timely may affect future business.

Ultimately, these costs need to be balanced carefully by the Owner and contractor, to not only stop at the immediate billable value of impacts, but to their consequences, that can reach further. When estimating the compensation, it shall also remain consistent and compared to other solutions altogether – acceleration, transfer of responsibility (de-scoping), etc. It is not rare that by asking for too high a compensation for a delay clearly caused by the Owner, the contractor loses the associated scope to a competitor as it was cheaper for the Owner to terminate that part of the contract and re-bid it.

Time Options, Windows and Costs of Opportunity

Time options and mobilization / construction windows also have a huge value for organizations. During contract negotiations, the determination of which party has the hand on the option or the window is an extremely important and valuable element. Its value increases the closer the Project comes to the date, in the form of a Cost of Opportunity. Project Managers need to understand this concept.

The value of a time option or mobilization window is linked to the indirect impacts of its application or its change. For example, if an expensive asset is to be mobilized on a certain date, and the Client wants to change the date, the value of that change depends on the capability to market the asset on other works. As this possibility decreases the closer one gets to the date, a last-minute change of date by the Client should lead to almost a full reimbursement of the asset availability. On the other hand, if the Contractor has the hand on the date, it give it a large flexibility to accommodate other opportunities for the asset, which then should justify a rebate on the price. As Time options and windows alter the opportunities to value the asset, changing them has thus a cost: the Cost of Opportunity.

Time options and window mechanisms will have been priced as part of the bargain that led to the initial contract. Any change needs to be considered in the light, not only of its direct cost, but of the related Cost of Opportunity. Project Managers need to be able to uphold these concepts when determining the value of claims related to changes of such dates. The concept can also apply for missing Owner deliverables that impede the start of certain operations, such as regulatory authorizations.

Hacking Through a Decision

Adhere to Scheduling Best Practices

Best practices of schedule management in respect of the contractual mechanisms are necessary to ensure difficult claims can be clearly expressed, understood and negotiated, or brought to court or arbitration with a positive outlook at recovery.

The focus of contractual claims will be on the Critical Path activities which determine the Project Completion Date. It is thus essential that as part of the development of the Integrated Project Schedule, the initial Critical Path has been properly identified. It needs to be consistent with the actual execution strategy, which might require some adjustments at schedule baselining stage.

As in all contractual matters, proper traceability and documentation of changes to the schedule, including updates and re-sequencing if carried out, needs to be ensured. This is achieved through the following:

- Systematic archiving of all subsequent updates of the schedule (generally, monthly updates), both in hard copy and in soft copy;
- Production of a monthly report tracing all changes of logic, activities removed and added, and any re-sequencing carried out with the explanations as to the source of the requirement for the change. This report should be produced by the planner(s), approved by a relevant authority within the Project, and archived with the schedule update. A summary version should be appended to the Project monthly internal and external reports.
- Should the Critical Path have changed due to changes in the durations of activity, this event should be raised immediately to the Project management and to the Client as this event has the

potential of creating claims, and will change the focus of the Project team.

Should there be a need for a full re-baseline of the schedule due a significant change in the conditions of Project execution, contracts generally stipulate that the re-baselined execution plan and schedule are to be submitted to the approval of the Client prior to their effective implementation, so as to avoid situations where the Contractor would use re-baselining to escape from contractual obligations by deeply changing the execution methodology that was agreed at the contract stage.

Issues with the Approval of Extensions of Time

As well as they are managed, however, those good practices are not sufficient to win most cases, as the main resistance to be overcome is human: a request for variation (cost or time) stays a request, enabled by Contract, but for which there is no pre-empted validation whatsoever, and many Owners are extremely reluctant in these matters. The reason is that they often have to refer to high level stakeholders and this might cause a personal reputation cost.

As a matter of fact, even if the delay event and its impact are clear (or simply arise from an instruction from the Owner), no contractual mechanism forces the Owner to promptly validate a request for Extension of Time, leaving many Projects in a state of wait, with the contractor on one side submitting for Extension of Time and an Owner on the other stalling on its award. The issue is that a change to the contract is not a unilateral decision: a consensus must be found. There will therefore be a need to fight for your case at management and sponsor level, after the logic argument, contractual entitlement and technical demonstration pre-requisites have been met. In this regard, the minimum is to have followed good practices, so that your case is consistent, well-structured and supported, as if it was a file meant for a submission to arbitration. The principles explained in the above Chapters shall be used to do so.

Negotiation Strategies

Three main strategies can be followed by the contractor for Extension of Time negotiations:

- Cooperative,
- Confrontational,
- Seeking independent external advice.

Cooperative Strategy

A forward-looking Project team will explore mitigation actions to disarm forecasted delay impacts - instead of having no other choice than to pay for the consequences of what has become a state of fact and often escalates into a dispute. Looking at constructive acceleration, extending for part of the scope only, proposing a re-evaluation scheme can all be pro-active solutions to enable the Owner to balance the expected Cost-Schedule impact, in line with Project objectives and contingency.

The contractor in this situation will be able to 'sell' more easily a situation which saves the resource it needs most (Time or Money), trading against the other, at everybody's benefit and avoiding reaching an impasse.

This strategy will in the end prove to be the most cost-effective for both parties. It however requires to have time and be reasonably convinced that both parties intend to be cooperative. A key requirement is to have pre-agreed shared objectives between Owner and contractor regarding the Project outcome.

In general, the cooperative strategy is rarely used when the size of the delays or associated compensation becomes extremely substantial.

Confrontational Strategy

A Contractor may be faced with a real need to obtain relief on time, and not a trade-off. In this case it will have to enforce its rights and insist on the formal approval of an Extension of Time. In this situation, and of course as allowed by the provisions of the Contract, his course of action would be of:

- not proceeding with instructions until a Change Order is signed,
- weighing in on his claims if they are not validated (highlighting and claiming cumulative impacts as time passes),
- being uncooperative (while staying within its obligations) on issues of great value to the other party,
- or even ultimately using suspension / termination clauses.

This course of action needs evidently to be supported by a strong and sound forensic analysis and argumentation. The damage to the relationship between parties should also be weighted in before choosing that course of action.

Using External Advice

In order to avoid lockdown situations which are detrimental to the Project, parties in disagreement also have the resort to ask for external review. A mediation, a third party review, or the submission of a subject to a Dispute Administration Board are all options that allow to get a neutral and purportedly fair eye on the dispute, its merits and the balanced risk of both Parties, with possible negotiated settlement or advice of solutions that had not previously been considered.

Conclusion

The schedule is a key instrument when it comes to claims related to delays (and related additional costs).

As a first priority, following the rules of proper Project scheduling exposed in this handbook and building a good quality Integrated Project Schedule with a clearly defined Critical Path will allow both Parties to have a sound and shared understanding of the Project dynamics. This should be accompanied by a flawless archiving of each schedule update and recording of all the changes implemented between each update.

Subsequently, a proper recording, documentation, traceability and notification of significant events between parties as soon as such events are known, are essential practices that will position the Contract party that will have best followed these recommendations in the best position in the case of a claim.

It is easier to take a decision on an Extension of Time when the event happens, because the impact and possible mitigations will be considered based on the information available at the time of actual decision. Forensic analysis after the fact is to be avoided if possible so as to avoid the controversial issues created by hindsight.

When it comes to claims, communication is key to make one's point. Using the Simplified Schedule to support to one's argument on the side of the detailed forensic demonstration is often very powerful. It is a scheduling presentation that can easily be understood by Senior Management.

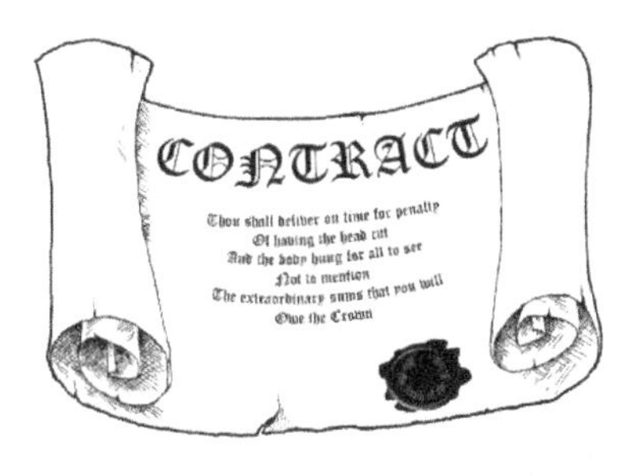

Chapter 13: Forensics of Project Schedule Management

Chapter Key Points:

- Failure Modes for the scheduling process follow the 3 key navigation questions which can be used as stages of maturity.
- Useful checklists are provided to check the schedule at different levels of process maturity.

This last Chapter will deal with some techniques to facilitate the analysis of Project Schedule management issues when undertaking a Project review in the midst of project execution. It is important to underline that:

- this Chapter mainly focuses on the issues related to the intrinsic quality of the scheduling process, not on issues related to the performance of the Project,
- the checks address Projects in execution stage only, as drivers in feasibility study/ tender stage will be different,
- the checks are for Project management and do not dwell into the detailed technicalities of scheduling.

Failure Modes for the Scheduling Process

As mentioned in the introduction, Schedule as a process needs to respond to the fundamental navigation questions:

- Where are we?
- Where are we heading to (if we continue according to the present trend)?
- What adjustments do we need to do to come back on course if we deviate?

Accordingly, schedule maturity can be classified at three different levels:

- Level 1 – Does the schedule reflect accurately the current position of the project and the progress of the works?
- Level 2 – Does the schedule include a proper forecast for the project completion that is relevant and resilient?
- Level 3 – Are the planners able to run scenario analysis, root cause analysis, and provide recommendations to the Project Management Team so as to take meaningful decisions?

This approach is relevant in our experience, as in many cases we encounter projects that fail the level 1 test, not having a proper updated schedule. Forecasting and scenario planning is then elusive without a proper basis.

Appendix 1 to 3 provides general checklists for Project schedules during Project execution that can be used on a regular basis and will not be repeated here.

Level 1 Checks – Schedule Update Accuracy

The first checks concern the basics of scheduling for a Project, as described in this handbook. The checklist below is to be complemented by the checklists in Appendix 1 and 3 on the quality of the Integrated Project Schedule.

Level 1 – Basics of Scheduling - Key Checkpoints	
Scheduling process Health Check (consider Appendix 1 as well)	
Does the Integrated Project Schedule cover all the project scope?	*Check with the Project team and the contract about scope that would not be covered in the schedule*
Have all changes to the scope been reflected in the schedule?	*Refer to the Project's Management of Change system (and Change Orders if any with Owner/ Client or suppliers) and check that approved changes have been reflected in the schedule where relevant.*
Have changes to the project execution strategy been reflected in the schedule?	*Changes in execution strategy can include changes in the procurement strategy, choice of locations for procurement or fabrication, etc. Identify those changes and check they are included with their consequences (e.g. transportation durations), to be consistent with the latest actual strategy.*
Are there reasons why certain situations / activities are willingly not shown in the schedule?	*Investigate for inaccuracies in the schedule that are consciously included due to stakeholder / contractual issues or other*
Have Budget Owners been designated for each section of the scope?	*Check that there is ownership of the delivery schedule for all engineering, procurement, fabrication and construction scopes*
Are Budget Owners satisfied with the durations reflected in the schedule?	*Check by sampling the consistency between what the schedule shows and what the Budget Owners believe is achievable*
(Construction phase only) Is the estimate of construction personnel on the ground consistent with what is shown in the schedule?	*Check by sampling the consistency between what the schedule shows and what the on-site Construction personnel believe is achievable*
Is the overall schedule owned by the Project Manager?	*How much use does the Project Manager of the schedule? Is it a reference document consulted on a daily basis to take decisions?*

Is the schedule's Critical Path consistent with what is identified by the PM as the Critical Path?	*Interview the PM to have his view on what is critical for the project. Check that this is what is reflected in the schedule (it can sometimes be very different, which is a problem)*
Is the Project Manager aware of embedded buffers?	*The PM should be able to designate embedded buffers in the schedule when they are not explicitly shown. Caution: if these buffers are on the Critical Path, the Critical Path is questionable.*
Has the schedule been properly weighted for progress aggregation?	*Check for the weightages allocated to the activities (it is fine if up to 25% of activities are not weighted and if weights are price-based for contractors (ref. Chap 4))*
Has the schedule been resourced? If yes, have the resource histograms been produced and mobilization rates been checked?	*Check that the best practices related to scheduling resourcing (Chapter 6) are followed: use of proven benchmarks, limitation of the number of key trades, resource density and mobilization limits.*
Schedule progress update	
Up to what cut-off point actual events and physical progress has been included in the schedule?	*Check what is the lag between the available information and what is shown in the schedule. Sometimes procurement, service contracts, fabrication and construction related information can be 1 or 2 months old depending on the reporting cycles.*
Has the schedule progress update been reviewed by the Process Owners?	*Before release of the updated schedule, the Process Owners would be expected to check for the consistency of the progress update even if it is generated from automated systems, or generated from a Third Party.*
Is the Integrated Project Schedule progress consistent with the Detailed Schedules/ progress?	*Check consistency over the schedule hierarchy, inasmuch as the Integrated Project Schedule being a summarized view, the transfer of progress data from Detailed Schedules might not always be straightforward*

Have all ongoing activities been actually progressed?	*On large complicated schedules it happens that the planners do not have time to update the entire schedule. Check by sampling that all ongoing activities have seen some reasonable progress compared to the latest update.*
Has there been instances where the progress had to be re-evaluated to a lower number?	*Check for any progress review that might have resulted in a lower progress than previously reported, which would show a flaw in the progress reporting process*
Against which baseline is the actual progress shown?	*If the project has been re-baselined, check which is the baseline used for reference*
Are there reasons why certain progress measures would be willingly not shown accurately in the schedule?	*Investigate for inaccuracies in the progress reporting that are consciously included due to stakeholder / contractual issues or other*
Is the progress measurement framework clear and consistent?	*Check the progress measurement method for each type of activity and whether it is implemented consistently (refer to Chapter 8)*
Re-sequencing	
Has there been some re-sequencing activities performed in the last period? If yes, have these changes clearly been traced in a log for future reference (e.g. in case of claim for EOT?)	*Re-sequencing can either be done at the request of the PM (when the schedule becomes problematic) or more stealthily by some Budget Owners or the planners themselves. Check for any re-sequencing and if there has been some, check that it has clearly been logged and traced.*

Level 2 Checks – Quality of Schedule Forecast

Level 2 checks are focused on the quality of the forecast. A forecast is only good if we know what is the current situation hence the level 1 checks must be positive before going into the level 2 checks which are slightly more elaborate.

Level 2 – Schedule Forecasting - Key Checkpoints	
Technical Schedule Reforecasting Issues	
Ongoing activities: has the reforecasting of the completion date been properly done taking into account actual progress?	• *Check by sampling that the end-date reforecasting takes into account actual productivity (is not just a push back by the amount of progress) and/or accounts for the opinion of the Budget Owner.* • *Check by sampling that for ongoing activities with no progress in previous period the end date is not just pushed by one period without analysis.*
Ongoing activities: upon update of the completion date, have linkages with the successors been changed?	*It might happen that activities take longer than expected but that there is enough substance to the deliverable to still start the following activity before total completion, whereas the initial plan was to wait for completion. These changes to the project logic are re-sequencing and should be properly registered and identified.*
Ongoing activities: for activities performed by Third Parties (suppliers, service contractors, fabricators etc.), has the forecast been validated by the Budget Owner?	*Review and validation of data provided by other parties is extremely important. Check that such a review has been done and that the project schedule reflects the view of the Budget Owner, not necessarily the view of the Third Party which can be excessively optimistic.*
Ongoing and past activities: has schedule productivity analysis been applied and data gathered that is useful for benchmarking and future similar activities?	*For substantial deviations from the initially planned duration, an analysis of the reasons of the deviation is expected. Sample those activities which duration appears to be substantially longer than the baseline and check if an analysis has been performed. A representative productivity measure is to be considered depending on the type of activity.*

Future (not started) activities: • has their start date been modified according to the schedule logic? • Has their duration been reviewed in light of observation of current or past activities?	• *Check by sampling that the schedule flows properly without undue constraint and thus that that start dates of future activities have been modified accordingly* • *Check by sampling that durations have been reviewed when information was available about actual schedule productivity in the Project conditions.*
Has the Project completion date been changed?	*Check whether the Project completion date (or essential delivery milestones) has changed and assess the issue.* *Check whether hidden buffers may have been used to maintain the project completion date*
Have implicit or explicit buffers been used to protect the Project completion date?	*Check whether implicit (hidden) or explicit buffers may have been used to maintain the project completion date. This is an indication of an increase of criticality*
Is a set of S-curves being produced at a relevant breakdown level? Have the trends been examined? Has there been reforecasting done as a result?	*S-curves should be produced at an adequate intermediate level (not at the Work Package and not for the entire project either). They should be analysed compared to the baseline for schedule productivity, and the current trend should be observed as well. It might be useful to also show the baseline's early and late curves for a further indication.*
Project reforecasting taking into account criticality	
Is the Critical Path the same after schedule reforecasting?	*Check the updated schedule for an actual Critical Path and whether it has changed*
How has the subcriticality of the other activity chains evolved?	*Check that the float on the other activity chains remains substantial.*
How has the overall float of the schedule evolved?	*It is possible to calculate an indicator of float over the entire schedule or parts of it (download the task and float data in Excel for the last few schedule updates)*

Has the Convergence Plan been updated and are there problematic deliverables for future gates?	*Deliverables that are forecast to be produced after the expected date in the Convergence Plan are indicators of potential criticality issues. They need to be investigated as soon as the potential for delay is identified.*
Is float monitoring implemented for critical deliverables and is the trend converging?	*Check float monitoring data. If such an analysis is not performed it is still possible to download in Excel from the latest 4-5 schedule updates raw files with tasks and float data, and analyse this data in Excel for float change (either the schedule's float or the float with respect to a fixed date)*

Level 3 Checks – Quality of Decision Support

Level 3 checks are only relevant when it happens that schedules are properly updated and reforecast. They simply aim to check that scenario planning is a capability that is embedded in the Project and used regularly.

Level 3 – Scenario planning- Key Checkpoints	
Is the PM involved in re-sequencing activities?	*Check if the PM is involved in re-sequencing reviews of the schedule to optimize execution*
Has the Simplified Project Schedule been updated regularly to fit with the latest schedule update?	*Check the status of the Simplified Project Schedule and whether it effectively reflects the current drivers of the project*
Does the PM have a need for scenario analysis?	*Check whether scenario analysis is a need for the PM and whether he is aware of the possibility to do those. Get his feedback on the easiness/ difficulty to get scenario planning results*
Has the planning group produced relevant scenario analysis at the request of the PM?	*Check on the planners side if they feel ready and tooled to provide relevant scenario planning data to the PM*
Is scenario analysis used as a support of re-baselining activities?	*Check at the last re-baseline how it has been conducted and if scenario analysis has indeed been used as a support for decision-making*

Conclusion

Schedule forensics sometimes requires significant persistence to really identify the issues at stake. Process-related sources of surprises can include basic Schedule update issues (level 1), or more advanced forecasting issues (level 2). The most frequent cause is poor structuring of the schedule hierarchy and poor build-up of the base Integrated Project Schedule.

If the schedule process has been implemented poorly it can require some work to reflect the actual situation and forecast of a project which can have a significant impact on the organization's commitments and thus results.

Schedule forensics must be undertaken by personnel experienced in Project Control knowing the particular business and that act independently of the project to shed the light on the actual causes of possible failures. In addition, the review must be sponsored by a person in the organization that is ready to listen to a view of reality that might contradict the official version, and act on this knowledge.

Conclusion: The Project Schedule, your Navigation Tool

All too often we observe how Projects start without having a proper navigation plan in hand. And then they poorly perform taking the sights and forecasting where they might land. There they float on the vast ocean, being more driven by winds and waves than driving towards their destination.

While in history this has sometimes, in very rare cases, led to astounding discoveries, like when Christopher Columbus landed in the Americas instead of the Indies as he expected, this is generally not what is expected in modern organizations executing Projects.

Sometimes we encounter these haggard and tired crews that only managed to reach civilization after long and varied adventures that all have the common point that at some stage, they were not any more able to regain control of their Project.

Project Value Delivery's mission is "to Empower Organizations to be more Reliably Successful in Executive Large, Complex Projects". It is in this spirit that this handbook has been designed: give Project management teams an appropriate framework to be more reliably successful, and remain in control, masters of their own fate.

Much of it can be summarized into designing proper navigation tools, and then implementing rigorous updating and forecasting processes with discipline.

Beyond this, a common thread of this book is to show how the collaboration of the entire team is essential to devise the best execution plan, and to integrate the information informally available throughout the team in a single repository that will serve to take decisions – for everybody's sake.

Scheduling cannot succeed in isolation. Its fundamental duty is to collect data from all stakeholders of the Project, bring it together in a consistent format and then revert with important coordination data. It relies in that respect on the cooperation of all functions and a far reaching communication capability with the entire Project team. Only this will enable the scheduling team to catch the right information required to produce an updated forecast reflecting the latest Project knowledge and position.

High quality Scheduling is a strategic competency for all international Project-driven companies. It is a differentiator and a competitive advantage in the market and industry. Poor scheduling has led Projects astray to the point of non-return as they lost control on what was happening. As per Project Value Delivery's mission, with your help and support, we will continue to strengthen scheduling as a discipline throughout Project execution organizations to make the execution of Large, Complex Projects more reliable.

Appendix 1: Integrated Project Schedule Health Checklist

Note: this checklist is only applicable during Project execution.

1	Did you define a clear schedule hierarchy?	
2	Does your Integrated Project Schedule typically have less than 2,000-2,500 activities?	
3	Is the ratio links/activities between 1.5 and 2.2?	
4	Have you been able to setup a sound update process for all Integrated Project Schedule activities?	
5	Is your schedule balanced in terms of activities between functions (in particular Engineering – Procurement – Construction – Commissioning)?	
6	Do your standard schedule printouts include: Total Finish Float column, baseline for all activities?	
7	Does your schedule show a clear Critical Path?	
8	Are the Total Finish Float values consistent with the expected schedule criticality of the Project?	
9	Is your schedule coded so as to allow views by function/ by area or deliverable/ scope/ Contractor (when relevant); each view on a number of levels.	

10	Did you define standard filters and views for all interested Project stakeholders?	
11	Did you minimize the number of constraints in the schedule?	
12	Did you check for negative lags in the schedule and did you replace them by positive lags?	
13	Did you check for Start-to-Finish links in the schedule and did you remove them as much as possible?	
14	Is your schedule network free-flowing (try to increase significantly some activities to check how the schedule reacts)?	
15	Did you resource the schedule for critical resources?	
16	Is the maximum level of resources consistent with the available resources?	
17	Is the maximum density of resources per work area consistent with work conditions (site congestion, personnel logistics)?	
18	Are the mobilization and demobilization curves for the critical resources realistic?	
19	How sub-critical are non-critical sequences of activities? Can you find a way to increase sub-criticality to increase schedule resilience?	
20	Identify the convergence points of activity chains on the Critical Path. How much float is available? What is the risk of concurrent criticality (merge bias). Is there a way to reduce this risk?	

Appendix 2: Monthly Schedule Update and Re-forecasting Checklist

1	Did you manage to effectively update all active activities with relevant input? Are there areas of the schedule that could not be updated this time? Was it due to lack of information or lack of time?	
2	Is the update in line with the opinion of the people on the worksite/ in the team? How did you check?	
3	Is the short term forecast in line with the opinion of the people on the worksite/ in the team?	
4	If the schedule shows some shift of the Project due to delays on critical activities, did you create buffers to constraint non-critical sequences of activities (so as to avoid "virtual float" creation)?	
5	How close is the Critical Path to switch because some other chain of activities' float is diminishing fast? What would be the implication of changing the Critical Path for the project?	
6	Did you analyse schedule productivity SPI(t) by main type of activity for those activities which progress reach 20-30%? Did you re-forecast accordingly?	

7	Based on the re-forecast of current activities, how is the timing of future related activities affected? Did they move as expected or would links be missing or inadequate?	
8	Based on the observed productivity of past and ongoing activities, and the knowledge acquired during this analysis, did you reforecast accordingly the duration of similar future activities?	
9	Have you updated the Convergence Plan without changing the gates' dates?	
10	Have you defined clear actions to catch up on future Convergence Plan gates that can be expected to be delivered late (this might require resources external to the project)	
11	Have you updated the Simplified Project Schedule in line with the Integrated Project Schedule update?	
12	Did you implement float monitoring on the main convergence points to analyse the actual convergence of the main sequences of events?	

Appendix 3: Project Lifecycle Success Factors

At Feasibility / Tender stage

1	Have the long lead items been clearly identified together with the relevant engineering, pre-award procurement activities and pre-manufacturing activities?	
2	Have specifically complex qualification and development activities been identified and included in the schedule with the relevant buffers?	
3	Has any constraints on the availability of a specific construction enabler been identified and included in the schedule?	
4	Has any issue and uncertainty related to soil conditions and preparation been identified and included in the schedule?	
5	Has a detailed construction schedule been developed that allows to cost the construction spread (sum of all equipment and manpower)	
6	Has a resource-based schedule exercise been carried out for those activities that might be resources-constrained in particular for construction? Are the rates of mobilization and demobilization reasonable? Has the level of resources been checked regarding congestion of the worksite, workers logistics and accommodation?	

7	Has there been any pressure to reduce the schedule compared to its natural duration?	
8	Has there been an analysis of the convergence points on the Critical Path and the actual sub-criticality of the other activity chains?	
9	Has a SSA been performed and the outcomes in terms of schedule resilience input into the Integrated Project Schedule?	

At Project Start-Up

1	Has a Convergence Plan been developed in the first 2-3 weeks of the project?	
2	Is the Integrated Project Schedule developed with particular emphasis on the interfaces between functions?	
3	Is the interface between Engineering and Procurement (main requisitions and scopes of work for service contracts, and vendor data for design finalization) properly represented?	
4	Is the interface between Procurement and Fabrication properly represented (material and equipment)?	
5	Have logistics durations been properly taken into account for material/ equipment delivery?	
6	Is there sufficient time allocated to Procurement pre-award activities, and award to effective start of manufacturing/ procurement?	
7	Is the interface between Engineering, Procurement and Construction properly represented (major installation aids)?	
8	Is the interface between Procurement & Fabrication, and Construction operations, properly represented?	

9	Is there a proper representation of the duration of Factory Acceptance Tests, Site Integration Tests and has the need for specific equipment been properly identified?	

At Engineering stage

1	Has a detailed Master Document Register been developed which dates are consistent with the overall Integrated Project Schedule?	
2	In case where the engineering process includes many inter-dependencies, has a detailed engineering schedule been developed, including resourcing per function to assess realism?	
3	On what assumptions have the relevant interfaces with Procurement been assumed (scope of work, technical bid evaluations, vendor data). Do these assumptions need to be revised in view of the market situation fed back by Procurement?	
4	Should there be any change in the Procurement, Fabrication and/ or Construction strategy as a result of engineering, has this been reflected in the schedule?	

At Procurement/ Fabrication stage

1	Have detailed schedules been requested from the suppliers of complicated equipment? Are they of the adequate quality?	
2	Have detailed schedules been requested from the fabricators? Are they of the adequate quality? Are they resourced in detail? Are they realistic?	

3	Have the relevant deliverables (free-issued items) to suppliers or fabricators been clearly identified and included in the suppliers' and fabricators' schedule, as well as in the Project's schedule?	

At Construction stage

1	Has a detailed (aggressive) Construction schedule been developed? Does it have the relevant interfaces with Procurement, Fabrication, service contracts (logistics)?	
2	Has a routine been setup for the generation of the detailed as-built schedule during Construction?	

At Commissioning/ handover stage

1	Has a detailed Commissioning schedule been developed? Does it have the relevant interfaces with Construction *(commissioning is by system and construction by area, hence the key deliverables need to be identified that are relevant to describe the interface)*?	
2	Has a routine been setup for the generation of the detailed as-built schedule during Commissioning?	

At Close-Out stage

1	Has an as-built schedule been produced, together with relevant schedule benchmarks for specific operations? (*this is applicable to the Integrated Project Schedule as well as the detailed Construction & Commissioning schedules*).	

Appendix 4: Ten Useful Schedule Rules of Thumb

Here are some useful, simple rules of thumb for the Project Management Team.

1. **Parkinson's law:** work tends to fill the time available to do it

   What can you do? *Make sure that activity durations remain challenging! (cf. the virtual float syndrome and the estimate padding syndrome)*

2. **Student's syndrome:** people will always start their tasks as late as possible

   What can you do? *Make sure that activity durations are short enough to avoid this effect!*

3. Much better have a very constrained and challenging project schedule with a buffer owned by the Project Manager than a relatively relaxed schedule where the owner of the float is not identified.

4. "An activity that is delayed by one month every month will never happen."

5. "No project plan will survive the first encounter with the bulk of reality." Remain agile.

6. If there are activities with Total Float over 100 days it is suspect for inappropriate linkage. Investigate.

7. Any Integrated Project Schedule that has more than 2,000 – 2,500 activities is suspect for over-complication and intrinsic mistaken links causing poor schedule update, disastrous re-forecasting and generally loss of control. Be cautious about the information and check the actual progress.

8. Schedules for complex projects are generally optimistic by 15-20% because they don't consider natural variation of activity durations.

9. People generally pad their estimated in particular if they are being asked to commit. Routinely ask people to perform cognitive work (e.g. engineering) in 50% to 70% of the time they announce as an 'achievable' target, and challenge severely all other announced durations to make them 'achievable' rather than 'quite sure'.

10. On a complex project, if you are late by T compared to your baseline schedule, even after a deep review, chances are that you are ultimately going to be late by at least 2 x T (proven by 'black swan' mathematics).

Appendix 5: Schedule Levels and Work Breakdown Structure

This Appendix aims at completing the explanations of Chapter 2 on schedule hierarchies and the need to have a set of detailed schedules by trade or function. It highlights the need of a consistency with the Project's Work Breakdown Structure when designing the finer grained schedule breakdown.

Bridging our Approach with Conventional Schedule Hierarchies

Many companies and management standards (including the AACEI and the PMBOK) define schedule levels (1, 2, 3, 4, 5) as common practice and define them as per their technical characteristics (ref. Chapter 2). In our consulting assignments, we advocate for a more purpose-led set of definitions. In this regard the Convergence planning is a level 1, the simplified project schedule a level 2, the integrated project schedule a level 3, and the detailed functional schedules a level 4. However, the common culture (often mentioned in contractual arrangements) is to have typically 5 schedule levels. It is generally the framework in the mind of scheduling and project professionals. While we are certain that the project culture will evolve over time, there is also a need to bridge the gap between technical schedule levels definitions and our objective-led definitions.

The key users, owners and typical schedules involved are summarized in the following table:

Schedule level	Key user	Owner	Typical schedules
Strategic	PM, PMT	Lead Planner	Level 0: convergence and objectives view Level 1: definition of means and windows (operating centers, constraining equipment mobilization, etc.)
Management	PMT	Planners	Level 2: Alert notices, EoT notifications, SSA Level 3: CPM, EoT analysis, manpower resourcing
Operational	Functional lead	Functional coordinator, site engineers	Level 4: lists, progress measurement, cycles fragnets Level 5: specific cycles/ hourly schedules

Many schedule practitioners fall in two common traps:

- Producing level 1 or 2 schedules without any intent behind them, as a simple roll-up of the lower level, simply in order to comply with a contract or management requirement,
- Failing to develop level 4 schedules, as they only are seen as a burden to entertain progress measurement, and left to crumble as soon as the management has their attention on another point.

This is the reason why we propose a slightly different approach in this handbook.

However we recognize that in many cases the framework within which most Projects are set impose to work by those standard schedule levels without defining their intent, and only their technical characteristics. By relating the scheduling levels to the WBS levels, some consistency can be brought into the scheduling hierarchy which will help produce a consistent set of schedules.

Fundamentals of Work Breakdown Structure Build-up

A schedule is nothing but an arrangement of activities in a logic network. The part not to miss out is the arrangement, which shall allow to sort out the scope, ensure nothing is missed out or doubled and everything can be found naturally.

The Work Breakdown Structure (WBS) is the standard control tool for comprehensiveness. A proper WBS allows structuring hierarchically the Project activities in order to assign Budget Owners, and handle the Project in manageable human-size sub-parts. The basis for proper planning is structure, which will also allow to setup the various control tools according to the same approach (progress measurement, material control, responsibilities, etc.). It is too often forgotten.

An effective WBS is a matrix between subjects and objects:

- A subject accounts for a trade, calling on specific knowledge, cycles, deliverables and tasks,
- An object is a physical part of a project.

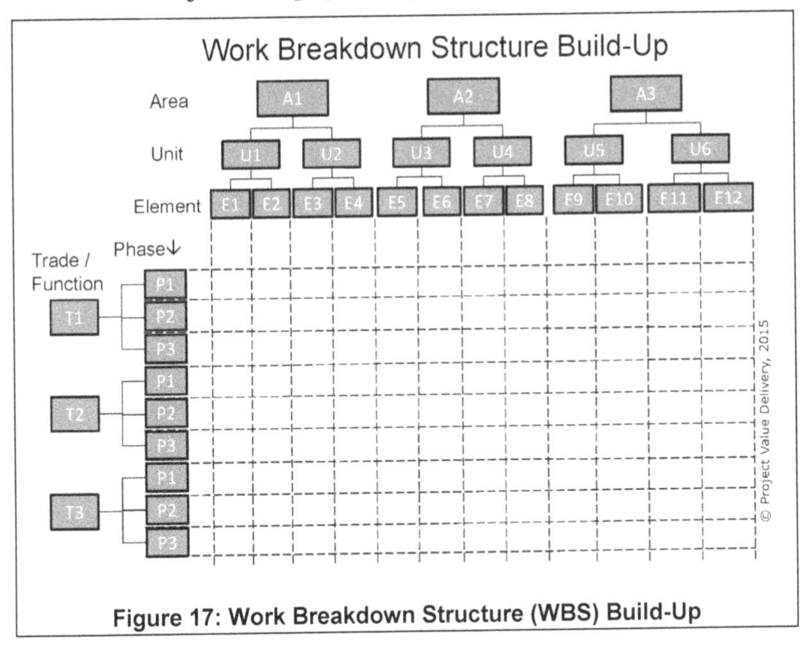

Figure 17: Work Breakdown Structure (WBS) Build-Up

Practically speaking, subjects are usually organized in two levels:

- Trade (electricity, process, civil, systems, etc.),
- Phase (Milestones, Engineering, Procurement, Construction, Commissioning, etc.).

Objects follow three levels depending of the size of the project:

- Area (plant, storage, utilities, off-site facilities, etc.),
- Unit (platform, process unit, etc.),
- Element (module, building, road, etc.).

The important concept is the matrix organization of physical objects with subjects, which will allow the splitting of tasks and activities in sub-entities of the project. The names for the levels of objects and subjects, and their list may of course be customized for each project.

It is extremely important to allow this structure to be built at the start of the project and followed thereafter for evolutions. It is central to the proper organization of activities and processes, avoiding the threats of having too low or too much detail. When it comes to schedule, this structure will drive how the planning is built, and how the different schedule levels are articulated. It has also a significant influence in the organization of physical progress measurement.

Producing a Schedule Hierarchy Consistent with WBS levels

For levels 0 through 3, a rule of definition is that the level of the schedule is representative of the same number of WBS levels plus one, and an additional level of detail for the represented tasks:

Level (Schedule) + 1 = Level (WBS)

- The Level 1 is either by Phase and Area or by Phase and Trade,
- The level 2 is either Phase / Area / Unit or Phase / Trade / Area,
- The level 3 is commonly by Phase / Trade / Area / Unit.

The levels 4 and 5 are more specific as they are only representative of a subsection of the entire Project, as a detail of the level 3. Level 4 generally represent the cut-out of one trade covering all areas, units and elements for a particular phase. Level 5 schedules will consider finer detail in terms of areas and units.

If we want to keep the level 3 schedule within a limit of 2,000 to 3,000 activities, the number of activities per Phase/ Trade/ Area/ Unit needs to be limited drastically. In a typical project there might be 3 phases, 5 trades, 20 areas/units. Then for each, only 5 activities are allowed on average. This shows the importance of more detailed level 4 and 5 schedules for driving the work of each trade.

One of the important and under-used tools are the cycles or fragnet, which are simply a stand-alone detailed schedule (level 4) of a repetitive task. It allows to set the logic and parameters of a sequence once, and then represent it simply at level 3 as a single task, with a duration proportional to its representative quantity. For example, cable laying is a sequence which detailing (cable tray installation, supporting, cable drums preparation, cable pulling, terminations) is of paramount importance and need to be detailed as a fragnet by a construction manager, to work with his subcontractors. Then it can be reduced at its simplest expression in a level 3 schedule: cable pulling of Unit A12, 10.112 linear meters, 26 days.

Conclusion

Notwithstanding the formal constraints on schedule hierarchy it is always possible to create a meaningful set of schedules. A key driver is to be consistent with the WBS when building that hierarchy, and keep the level 3 Integrated Project Schedule within a reasonable level of detail.

Appendix 6: Earned Schedule Management

Earned Value Management (EVM) is taught as the key forecasting tool in most Project management courses. In practice it is a too theoretical approach for most Projects and can rarely be used in a straightforward manner. Still, in can provide the Project Manager with a useful analysis tool that can be leveraged during the forecasting exercise. Because EVM converts all data into $ value, it can also deal with the aggregation of quite diverse activities.

In this appendix we explain a specific version of EVM with a focus on schedule analysis and forecast: Earned Schedule (ES) . Earned Schedule has been formalized in 2004 by Walter Lipke and further developed in his book 'Earned Schedule'. It is still not accepted as a mainstream tool, but does make some inroads into codes of practice. Conceptually it is a simple development of EVM and appears much more stable, in particular when Projects are subject to substantial delays compared to the baseline. We have chosen to use ES as a reference in this handbook because if its practicality.

Earned Schedule must be applied on a set of activities that is consistent in terms of drivers and resources. The application at a low level such as the Work Package level or some consistent subset of the WBS is thus the most relevant.

Prerequisites of Using Earned Schedule

Contrary to Earned Value, Earned Schedule does not require alignment between cost breakdown structure and schedule breakdown structure. It is sufficient to have a properly weighted schedule. However, if Earned Value management is expected to be used for cost analysis and forecast at the same time, such alignment is required – refer to our Cost Control Handbook.

Objectives of Earned Schedule

At any point during the Project, Project Managers must have access to the information that allows them to answer two basic questions:

- Is the Work Package currently ahead or behind schedule and, if so, by how much? (we will refer to this as determining the Project schedule variance.)
- If the project continues according to the current trends, when will it be completed (determining the schedule completion date forecast)?

Conventional Earned Value (EVM) Basics

For these questions to be meaningful at all, a Project Manager must first have a baseline plan that defines where the Work Package was expected to be in terms of schedule and costs at any particular point in time. This plan can be summarized as the Work Package time-phased budget, or budget baseline. On the figure below we have taken what we have referred to so far as budgeted costs and replace that name with a more precise term — "Planned Value", or PV. The budget baseline allows the Project Manager to determine the budgeted cost of the Work Package at any point in time.

In the following illustration, at the current Actual Time (AT) we see a situation in which the Work Package progress appears to be lower than what was expected.

The planned cost of work actually performed is referred to as the "Earned Value" or EV for the Work Package. We can use this curve to determine the schedule variance by comparing the budgeted cost of the work we planned to have accomplished at the current time to the budgeted cost of the work that was actually performed (Schedule Variance (SV) = EV-PV).

The conventional EVM schedule variance SV is computed using as the weight for each activity the budget cost (or price), therefore the schedule variance is only the summation of physical progress over the entire work package activities without considering any Actual Cost overrun effect. This is an approximation which is sufficient in most cases.

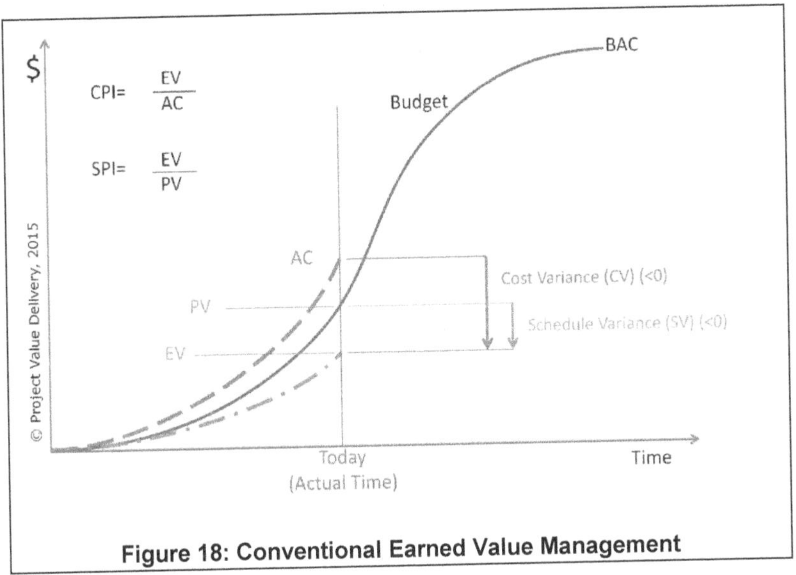

Figure 18: Conventional Earned Value Management

The figure shows that the Work Package has an unfavourable schedule variance, that is, we are currently behind schedule. Using this convention, a favourable schedule variance is represented by a positive value, while a negative value represents an unfavourable variance.

We can also represent on the same curve the Actual Cost of the work performed so far. The Actual Cost of work performed (AC) is what has been spent to date. In the example below, it exceeds the amount budgeted for that same work (AC > EV). In other words, our Work Package is showing an unfavourable cost variance — a cost overrun.

In traditional cost/schedule systems, the cost variance is determined by subtracting AC from EV (Cost Variance = EV-AC). (As in the case of schedule variances, favourable schedule variances are represented as positive values, while unfavourable variances produce negative values).

Alternatively to the Cost Variance and Schedule Variance, the Cost Performance Index (CPI) and Schedule

Performance Index can be used. If the index is above 1, it is a favourable variance; if below 1 it is unfavourable.

$$SPI = \frac{EV}{PV} \quad \text{and} \quad CPI = \frac{EV}{AC}$$

These indexes are actually productivity factors (either for cost or schedule).

These ratios can then be used for forecasting. The 'optimistic' forecast considers that whatever happened before, the future will happen as per the initial plan in terms of productivity. The 'pessimistic' forecast will apply the same productivity factor for the future than what has been measured in the past.

New Project Management Institute (PMI) terminology

ACWP, BCWS, BCWP are now obsolete terminology for the Project Management Institute. If you plan to pass your Project Management Professional (PMP®) certification, note that these terms have been replaced:

- ACWP (Actual Cost of Work Performed) by AC (Actual Cost)
- BCWS (Budgeted Cost of Work Scheduled) by PV (Planned Value)
- BCWP (Budgeted Cost of Work Performed) by EV (Earned Value)

Earned Schedule

While Earned Value Management works very well in the case of cost, it is more difficult to apply in the field of schedule, for the following reasons:

- The schedule variance SV is measured in the same units as the cost variance, monetary units such as dollars. This is not easy to translate into actual tangible time delays,
- The schedule productivity ratio SPI is proven to be unstable in particular for projects that get significantly delayed – after the initial completion date is passed, SPI will converge back to 1 and will not really measure the

actual schedule productivity. It must thus be used with care.

Usage of conventional SV and SPI is thus not recommended. Earned Schedule intends to remedy these drawbacks by introducing a more straightforward approach. A new measurement is introduced, which is called the 'Earned Schedule' (ES). The Earned Schedule is what the time in the project should have been as per the initial plan to reach the current progress (or Earned Value). The concept is shown on the following figure.

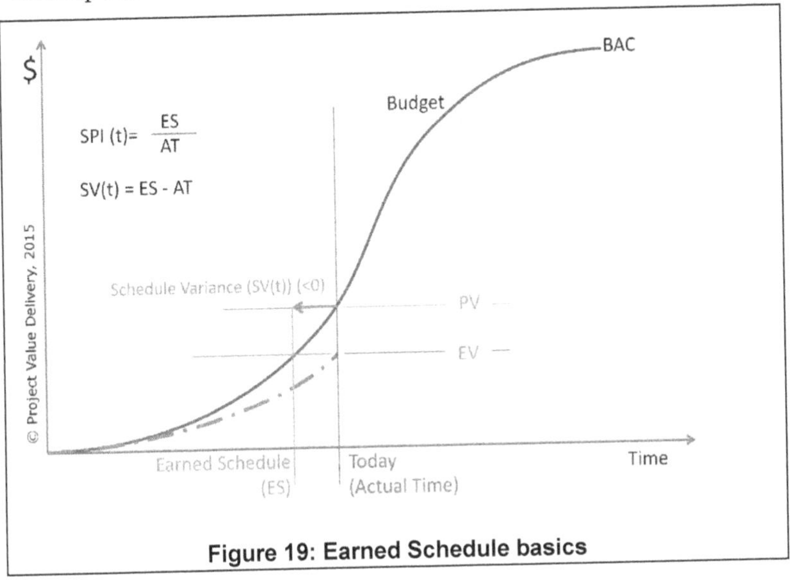

Figure 19: Earned Schedule basics

New schedule variance and productivity measurements can then be introduced. The Schedule Variance SV(t) is the difference between the Earned Schedule and the Actual Time: SV(t)=ES-AT, expressed in time units, while a new schedule productivity ratio is

$$SPI(t) = \frac{ES}{AT}$$

For example, in practical terms, let us suppose that the Actual Time is 6 months into the project. We have earned a value of 10,000$; and we were supposed to have earned such an amount only 4 months into the project as per the initial plan, ES=4 months, and SV(t)=-2 months, SPI(t)=6/4=1.5.

We have presented Earned Schedule under the same format as Earned Value to show familiar graphs and compare the two methods. In reality we do not need to measure the Earned Value in monetary terms to apply Earned Schedule. Properly weighted percentage progress is sufficient and adequate. To use the same example (supposing the initial budget is 40,000$), we can write the same sentence as: Actual Time is 6 months into the project. We have earned 25% progress and we were supposed to have earned such an amount only 4 months into the project as per the initial plan, ES=4 months, and SV(t)=-2 months, SPI(t)=6/4=1.5. Hence, the Earned Schedule method does not require any monetary valuation of Earned Value – as long as the progress measurement framework is consistent and properly weighted.

Studies have shown that these measures SV(t) and SPI(t) are more representative of actual schedule productivity, and they are also more stable during project execution than the conventional Earned Value Management measures. They should be preferred when it comes to measure schedule productivity.

Using Earned Schedule for Forecasting

Once you have evaluated the Schedule Variance of the Work Package to determine its status, the next question is "what will be the final duration of my Work Package?"

As in cost Earned Value, there are different ways to forecast.

- The optimistic way is to consider that anything that hampered progress in the past is resolved and that the future activities will be as per the initially planned productivity. The final delay of the project will be the current delay.
- The pessimistic way is to apply the current measured schedule productivity SPI(t) to future activities as well, forecasting the completion date accordingly.

Studies of hundreds of Projects have shown that Project performance rarely tends to improve. In fact, once a Project passes the 15%-20% completion point, performance almost never surpasses the average performance to date, and often gets worse. Exceptions concern cases where specific issues impeded the start of the project, which have safely been removed.

As a minimum, the observed productivity factor, which is in fact the Schedule Performance Index SPI(t), should thus be applied to the future work to build a forecast of the completion of the Work Package. However, this view is generally only valid for the portion of the progress between 20 and 80%. The duration of the final progress beyond 80% must be considered separately, as the productivity factor might not be a good guidance.

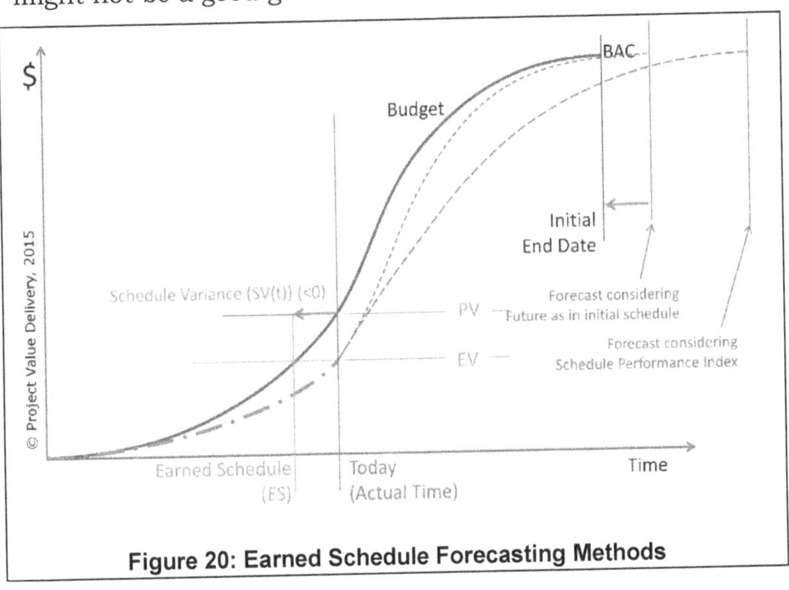

Figure 20: Earned Schedule Forecasting Methods

For the complete Project, the activities on the Critical Path have more importance than others. Hence Earned Schedule must be applied on the Critical Path activities; as well as on sub-Critical Path activities to check that they won't become critical. Usage of Earned Schedule is thus more difficult and needs to be considered with care.

These Earned-Value based forecasts provide a broad guidance and a rough forecasting model that can give useful orders of magnitude, in particular to challenge Budget Owners with observed productivity factors compared to the initial plan. However, whenever possible, reference to the actual root causes and issues faced by the Work Package will give a better insight into the mechanisms for performance and will result in a more reliable forecast. In all cases, it is from the observation of the actual issues faced that the Budget Owner will be able to revise its forecast.

Abbreviations and Glossary

AC (Actual Cost)	The actual cost that the Project has incurred to date. It was previously called ACWP (Actual Cost of Work Performed)
AT	Actual Time (terminology of Earned Schedule). It is simply the current time in the project since its effective start.
BAC (Budget At Completion)	The As Sold Budget plus any Approved Changes
Budget Owner	A designated manager within the Project team that is responsible for the delivery of a scope including cost and schedule.
Buffer	In this book, a duration that is under management control to constraint parts of the Project schedule and/or to protect completion dates. Buffers are represented by specific mock activities in the Project schedule, and their duration is under the control of management.
Contracts for Services	In this book, Contracts for Services designates any procurement of services (to be distinguished from Purchase Orders used for material and equipment). In most Contractor organizations, the terminology "Subcontracts" will be used.
CPI	Cost Performance Index, used in Earned Value Management to compute the cost productivity. CPI < 1 indicates a lower than expected cost productivity.
CPM	Critical Path Management method (conventional scheduling method)

Critical Chain	A development of Critical Path method under the Theory of Constraints (Goldratt). It takes into account in addition the resource constraints to determine what is the set of activities that effectively constraints Project delivery. A Critical Chain might not be logically linked if it is driven by the availability of scarce resources. Critical Chain requires specific software and resource-loaded schedules to be determined.
Critical Path	The sequence of Project network activities which add up to the longest overall duration. This determines the shortest time possible to complete the Project. *Note: the Critical Path activities have a float of 0 only if there are no constraints imposed on the Project network. Hence a definition via zero float values is not universally valid.*
CTR	Cost, Time, Resource (used for scoping engineering groups of tasks)
CV	Cost Variance in Earned Value Management.
DPR	Daily Progress Report
EoT	Extension of Time. An extension of the Project Completion Date granted by the Owner/ Client based on justification for the origin of the delay.
ES (Earned Schedule)	Earned Schedule, a variance of Earned Value Management for schedule.
EV (Earned Value)	The amount of BAC that has been earned. In the simplest form EV= % Complete * Budget at Completion. It was previously called BCWP (Budgeted Cost of Work Performed)
EVM	Earned Value Management

Float	Amount of time an activity in the Project schedule can be delayed without impacting either the subsequent activities (free float) or the Project completion date (total float). In this book, Float refers to total float for the sequence of activity under consideration. Float is a value that is calculated, but is not represented physically in the Project schedule (contrary to Buffers).
Fragnet	A stand-alone detailed schedule of a repetitive task, which allows this set of tasks to be represented by a single activity in the Integrated Project Schedule.
Function (or Trade)	Organizational department with a specific speciality. E.g. engineering departments, construction departments, procurement, etc.
Lag	A delay that is inserted as part of a logical relationship between two tasks. Example: *Task B starts 5 days after task A is finished. This is a 5 days lag on a FS logical relationship.* Lags should generally be avoided in Integrated Project Schedules and replaced by actual tasks. They are acceptable in high level schedule development to make the schedule development easier.
Liquidated Damages (LD)	A pre-defined estimate of the losses to the Client due to delays in the completion of the Project. The Contractor will lose the LD in case of late delivery. It is generally an amount per day of delay, up to a certain cap which is a significant percentage of the total Project value. The client is not entitled to claim more for late delivery, but application of LD's can significantly decrease the profitability of a Project for a Contractor.

Logical Relationships	Logical relationships between tasks are implemented in a schedule. The different types include: • FS Finish-to-Start (most common) • SS Start-to-Start • SF Start to Finish (to be avoided and forbidden for SSA) • FF Finish-to-Finish
MDR	Master Document Register
PCM	Project Control Manager
PM	Project Manager
PMT	Project Management Team – Project Manager and its direct reports, generally consists of 10-15 person on large Projects.
PPR	Project Periodic Report
Predecessor	A task that is logically linked to immediately precede the task under consideration (FS or SS links).
PV (Planned Value)	The amount of budget that would have been spent to date according to the baseline plan. Alternatively the amount of BAC that the Project would have consumed to date. It was previously called BCWS (Budgeted Cost of Work Scheduled)
RACI	Responsibility Accountability Consult Inform matrix. An effective tool to show responsibilities for different processes and deliverables.
Sensitivity	How the final estimated result could vary. It generally involves giving values for a best case, an achievable case and a worst case.
SPI	Schedule Performance Index, used in Earned Value Management to compute the schedule productivity. This ratio is not very reliable. Prefer to use SPI(t).

SPI(t)	Schedule Performance Index, used in Earned Schedule to measure schedule productivity. SPI(t)=ES/AT. SPI(t)<1 indicates a lower than expected schedule productivity.
SV	Schedule Variance in Earned Value Management, expressed in monetary terms.
SV(t)	Schedule Variance in Earned Schedule, expressed in time units
SSA	Schedule Statistical Analysis
Subcontracts	Contracts for services. The term is generally used by Contractors. In this book the terminology “Contract for services” is used.
Successor	A task that is logically linked to immediately follow the task under consideration (FS or FF links).
Virtual Float	Float that is created on a sequence of activities by delays on other sequences in the project. This creates additional slack that will be used by project contributors. Creating unchecked virtual float should be avoided.
Visual Management	A management method using visual dashboard to help coordinate the work of the team.
WBS	Work Breakdown Structure
WP	Work Package, the elementary breakdown used for cost and schedule control.

Table of Figures

Acknowledgments

From Both Authors

Many people have taken the time to review and comment earlier drafts of this version of the handbook, amongst which Jean-Pierre Capron, Johann Declercq, Vijay Gopinathan, Ingvar Skogland, Babu Surendran, Anthony Nouveliere. Many thanks to them all - as well for the great conversations on the topic of scheduling in general, and projects' execution in particular.

Companies in general, and consulting companies in particular, are shaped by their clients. In this instance, we would like to thank particularly Project Value Delivery's long-standing clients for their support and input in giving us the inspiration and drive to produce this handbook. We would like to thank in particular Subsea 7 for which an earlier version of this handbook has been customized, which process had led to long and interesting discussions. Our thanks go in particular to Allen Leatt and Antoine Guillard at Subsea 7. Other clients such as SBM Offshore, EMAS AMC, Eramet Weda Bay Nickel / Technip, Petrofac, SapuraKencana, SapuraAcergy, have also provided very worthwhile inspirations in different contexts.

All mistakes and oversights are the responsibility of the authors only. Corrections, suggestions or feedback should be sent to contact@ProjectValueDelivery.com for inclusion in subsequent editions.

From Jeremie

What would be an author without his family? Warm gratitude go to my wife and children who have endured the torments of the writer during the production of the series of books on Project Control.

A tip of the hat to all clients and colleagues with whom I have worked in the past twenty years for all the great conversations and insights into schedule management – sometimes when they did not even know that what we discussed would end up in a book!

Finally a big thank you from Thierry Linares who has significantly contributed to improve the manuscript of the book, in particular by rewriting Chapter 12 and Appendix 5.

From Thierry

This being my first paperback publication, I first need to thank Jeremie for this opportunity given to me to be his co-author, as well as the good moments when we regularly meet around the world.

I want then to express my deepest gratitude to my loving wife for her support and encouragements in this writing process, which began with white papers and my thesis, and to my children, for their patience.

I must particularly recognize my co-workers, management and staff of the last twelve years, with whom I experienced the issues discussed in this book and found the will to overcome them.

Finally, my appreciation goes to my employer, Technip and its management, for their confidence and the latitude entrusted in me to manage and find innovative approaches to project challenges.

Index

Project Value Delivery,

a Leading International Consultancy for Large, Complex Projects

This cutting-edge Project management book is sponsored by Project Value Delivery, a leading international consultancy that "**Empowers Organizations to be Reliably Successful in Executing Large, Complex Projects**".

Part of our mission is to identify and spread the world-class practices that define consistent success for Project leadership. Ultimately, we want to be able to deliver a framework that makes Large, Complex Projects a reliable endeavor.

Our Book Series are a crucial part of this framework, spreading indispensable good practices and skillsets for leaders in Projects.

Our approach to Project success

At Project Value Delivery we believe that Project success is based on three main pillars which require specific sets of skills and methodologies specific to Large, Complex Projects. All three need to be strong to allow for ultimate success:

- Project Soft Power™ (the human side)
- Systems
- Processes

We focus on embedding these skills and methodologies in organizations through consulting, coaching and training appointments. We develop what organizations need and then help them implement it sustainably, transferring the knowledge and skills.

We recognize that to be effective, our interventions will involve access to confidential business information and make it a point to treat all information provided to us with the utmost confidentiality and integrity.

Our Products

Our products are directly related to our three pillars. We have developed proprietary methods and tools to deliver the results that are needed for Large, Complex Projects. In a number of areas, they are significantly different from those conventional Project management tools used for simpler Projects.

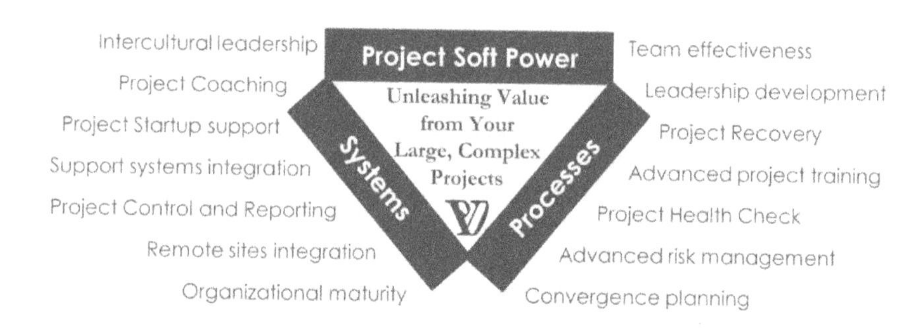

We focus on consulting, coaching and training interventions where we come in for a short to medium duration, analyze the situation, develop customized tools if needed, and transfer skills and methods to our clients so that they can implement them in a sustainable manner.

Contact

Contact us to learn more and get great resources for free!

Contact @ ProjectValueDelivery.com, and visit our website **www.ProjectValueDelivery.com** where you can register to receive regular updates on our White Papers.